Praise for *The Tyranny of E-mail* by John Freeman

"A thoughtful and provocative book."

—*The Seattle Times*

"An elegant self-help book . . . Freeman uses lush prose and invokes examples from great literature to make his points. He comes at things not from a giddy utopian perspective that permeates most writing about technology but from a humanist one. It makes the book refreshing and powerful."

—*The Boston Globe*

"[Freeman] brings the reader a fresh, intelligent look at e-mail's infiltration into and influence over every aspect of twenty-first-century life. . . . *The Tyranny of E-mail* serves as an engaging reality check."

—*The Daily Beast*

"Freeman offers up fascinating trivia . . . [and] makes a persuasive case that e-mail has at once corroded epistolary communication and strangled workplace productivity."

—*The New Yorker*

"E-mail is eating us alive . . . Luckily for us [John Freeman] has a solution."

—*Chicago Tribune*

"A book with a title this bold and provocative . . . requires an airtight and compelling case to back it up. To keep us reading, the book must also inform and entertain. John Freeman . . . delivers on all counts."

—*The Oregonian*

"Freeman vividly illustrates the moral, physical, and psychic costs of 24/7 availability."

—*Star Tribune*

"In an entertaining, easy-to-read style, Freeman details what has happened ever since communication moved beyond face-to-face conversation. . . . This book is a fascinating mixture of philosophy, psychology, history, sociology, electronics, and economics."

—*St. Louis Post-Dispatch*

"In prose both wise and entertaining, a book has arrived to awaken a nation of inveterate e-mail checkers from their collective lunacy. . . . *The Tyranny of E-mail* is a funny read filled with anecdotes you'll want to share with friends."

—*The Kansas City Star*

"A truly fascinating journey . . . [Freeman's] book is informative, well-written and researched, and certainly a wake-up call telling us that our technological lifeblood is more than a little poisoned."

—*The Grand Rapids Press*

"World-class book critic Freeman critiques the dark side of our e-mail habit within a penetrating history of written communication and how it has shaped society. . . . His love of language and literature is evident throughout, but Freeman is especially expressive in his musings about how disconnected we are from the living world as we focus on the digital realm."

—*Booklist*

"Freeman, acting editor of *Granta* magazine, captures viscerally 'the buzzing, humming megalopolis' that 'tunes into this techno-rave of send and receive, send and receive.' And he draws effectively on psychological and social research to describe the harm this 'tsunami' of e-mail is causing: fragmenting our days, fracturing our concentration, diverting us from other sources of information and face-to-face encounters. His closing 'manifesto for a slow communication movement' could fuel an e-mail rebellion."

—*Publishers Weekly*

"A cool yet unreserved manifesto against the drug-like, numbing consequences of e-mail correspondence and the alienating paradox at the heart of our so-called connected online lives. The author's ten-part program to break what he considers a dangerous addiction to e-mail is predicated on a simple, potentially divisive premise: Don't send one. Freeman is a matchless writer with a talent for such bombshells, and his conclusions are rational, practical, and wholly refreshing. First steps toward a slow communication movement?"

—*Kirkus Reviews*

"We live in a culture devoted to technology, and yet most of us cannot find the time to consider its history or its consequences. John Freeman has made the time, and has thought carefully about how we have gotten here. . . . Freeman knows his history, and he offers an engaging account of the evolution of correspondence."

—*Bookforum*

"A few decades ago, the ruler of Yemen ordered the northern gates of his city permanently barred, proclaiming, 'Nothing but evil comes through here.' After reading *The Tyranny of E-mail,* I'm feeling the same way about my laptop. Freeman's impeccably researched, eloquently argued book reveals the many ways this so-called boon to communication and productivity has become a distracting, privacy-sapping, alienating, addicting time-suck. He has convinced me that the new mantra for our times ought to be Tune out, Turn off, Unplug."

—Geraldine Brooks, Pulitzer Prize–winning author of *March* and *People of the Book*

"A mix of cool historical analysis and articulate outrage. Freeman doesn't just diagnose our creeping e-madness—he offers a cure that may prove liberating."

—Jim Holt, author of *Stop Me If You've Heard This: A History and Philosophy of Jokes*

"John Freeman brilliantly explores the paradox that increasingly defines our lives: the more we 'connect' through the Internet, the more disconnected we become. Closely argued and historically informed, *The Tyranny of E-mail* couldn't be more timely."

—James Shapiro, professor of English, Columbia University, author of *A Year in the Life of William Shakespeare*

THE TYRANNY OF E-MAIL

The Four-Thousand-Year Journey to Your Inbox

JOHN FREEMAN

Scribner
New York London Toronto Sydney

Scribner
A Division of Simon & Schuster, Inc.
1230 Avenue of the Americas
New York, NY 10020

First Scribner trade paperback edition January 2011

Designed by Carla Jayne Jones

Manufactured in the United States of America

1 3 5 7 9 10 8 6 4 2

Library of Congress Control Number: 2009032087

ISBN 978-1-4165-7673-0
ISBN 978-1-4165-7674-7 (pbk)
ISBN 978-1-4165-8812-2 (ebook)

Photographs on page 1 courtesy of Museum of Archeology, Istanbul; image on page 10 courtesy of the Print and Picture Collection, Free Library of Philadelphia; photograph on page 14 © ThinkStock/Superstock; photograph on page 43 courtesy of Smithsonian Institution, National Museum of American History; photograph of H.L. Mencken on page 54 courtesy of Yale Collection of American Literature, Beinecke Rare Book and Manuscript Library; photograph on page 62 courtesy of Western Union Telegraph Company Records, Archives Center, National Museum of American History, Behring Center, Smithsonian Institution; photograph on page 70 courtesy of Photography Collection, Miriam & Ira D. Wallach Division of Art, Prints & Photographs, The New York Public Library, Astor, Lenox and Tilden Foundations; map on page 88 courtesy of Bolt, Beranek & Newman; photograph on page 92 courtesy of Leonard Kleinrock; photograph of William Burroughs on page 103 taken by Brion Gysin, reprinted with permission of The Wylie Agency LLC on behalf of The Estate of William Burroughs; image on page 125 courtesy of Aubrey Jones; photograph on page 190 © Ingram Publishing/SuperStock; image on page 193: Samuel Yates, Untitled (Minuet in MG), 1999, mixed media, 15in. x 26in. x 65ft., Collection of Rene & Veronica di Rosa Foundation, Napa, CA. Photo by Ben Blackwell.

This book is for my grandmother,
who wrote the most wonderful letters,
and for my mother, who taught me how to reply

No man can be turned into a permanent machine. . . .
Immediately the dead weight of authority is lifted from his head,
he begins to function normally.

—Mahatma Gandhi

CONTENTS

INTRODUCTION

The oldest love poem in the world sits behind a glass case at the Museum of the Ancient Orient in Istanbul, where it was placed on display on Valentine's Day 2006. Carved in cuneiform, it rests on a clay tablet the size of a piece of toast, the script as small as bird tracks. "Bridegroom, you have taken pleasure of me," the poet, a ghost lost to time, pleads in Sumerian. "You have captivated me: let me stand trembling before you."

Love may not be forever, but this expression of it has outlasted swords forged by fire, cities designed by the finest architects, the largest machine ever to fly, and the most titanic boat ever to sail. To write his verse, the poet would have had to compose the lines in his head or recite them to a friend. Then he would have molded the clay tablet and slowly, but deliberately, carved his verse into it with a reed staff before the clay hardened. Finally, he would have dried the poem in the sun and waited another day for it to cool, when it could be delivered to his beloved by hand.

Feelings may not have a terminal velocity, but it should be said that certain expressions benefit from careful deliberation. Love is certainly one of them, but so is regret. So are longing, forgiveness, curiosity, and anger. Communication—the conveyance of meaning from one person to the next—depends on how we frame it. The second-most important question we must face, after choosing to communicate at all, becomes how to deliver what we want to say. Four thousand years after this poet bent over his writing desk, we have as many options as we have languages.

You can write your message in the sky, send it by singing telegram, speak into voice mail, shove it in the post, and hope for the best. You can write it in free verse, broadcast it to three hundred of your closest friends on Facebook, fire off an instant message, post it to your Twitter channel. If we think of modes of communication as a mirror spectrum of the human voice, we have as many registers as our mouths can make. The telex machine may have died, but most copy shops and offices still have fax machines. Phone booths still huddle, in various states of molestation, on many street corners. We can sign a message, pantomime it, text it, shoot a video message, record it as a song, upload a declaration of love onto YouTube, chalk it on pavement, scratch it on a tree trunk.

In his book *The Gift,* Lewis Hyde argues that one of the most effective ways to send a message into the world is to wrap it in a form that only it can possess and give it away. Why buy a card when you can make one? Why sermonize when you can write a sonnet? But how many of us have the time for this—or the skill? All over the world we are working longer hours than ever, sleeping fewer winks, taking shorter vacations. In this environment, frazzled and fried, tied to a machine that gazes back at us more hours per day than even our spouses do, we do what makes the most sense: we send our messages the fastest way possible.

The Inbox of Kings

In June 2004, the Internet giant Google made an announcement that quietly marked the apotheosis of the e-mail age. Gmail, its Web-based mail program, would offer users unlimited storage. Imagine for a moment what this means. Thanks to a group of 450,000 machines scattered across the United States like underground missile bunkers, I could store more e-mails than there are blades of grass in Kansas. This is beyond unprecedented—it is *superhuman.* Is God's inbox this big? Prior to the electronic age, dictators and kings did not enjoy such epistolary armories.

Still, their capacity is dwarfed by the Herculean arms of an everyday individual's e-mail inbox today. What busy individual needs this industrial-strength capability for his correspondence tool? What buzzing, humming megalopolis tunes in to this techno-rave of send and receive, send and receive? Is the human brain wired to receive this much stimuli? Can our eyes scan this many separate pieces of information? Is anyone listening? Who is it behind the screens, tapping the bellows and pumping the organ keys of this huge, throbbing machine at all hours of the night?

For the Love of E-mail

The answer, of course, is us. We love e-mail. In 2007, 35 trillion messages shot back and forth between the world's 1 billion PCs; in the time it took you to read to this point, some 300 million e-mails were sent and received. They sluiced down corporate drainpipes, piled onto listservs, promising a return on investment in a small African country and providing jokes about pigs and news about your grandmother's heart surgery. According to a Stanford University survey, 90 percent of all Internet users e-mail. In 2009, it has been estimated, the average corporate worker will spend more than 40 percent of his or her day sending and receiving some two hundred messages. Instead of walking down the hall, picking up the phone, or sending an interoffice memo, we e-mail.

E-mail goes with us everywhere now. We check it on the subway, we check it in the bath. We check it before bed and upon waking up. We check it even in midconversation, blithely assuming that no one will notice. We check from our loved ones' deathbeds. Even the most powerful people in the world do it. On most days during the 2008 presidential race, Barack Obama's BlackBerry "was fastened to his belt—to provide a singular conduit to the outside world as the bubble around him grew tighter and tighter throughout his campaign." President George W. Bush, who received fifteen thousand e-mails a day at the White House, said that one thing he looked forward to after leaving office was e-mailing. There is even a service that allows you to send an e-mail after you're dead. If there is an hour or a minute or a second to spare, e-mail is there. It is our electronic fidget.

It's hard to blame us. Once broadband connection arrived, e-mail became the world's most convenient communication tool. Not much more than a dozen years ago, most of us printed letters out, placed them in envelopes, and then walked or drove them to the post office, where we waited in line, wasting more time, so that the letter could arrive in maybe a week. The U.S. Postal Service estimated that, even if 99.8 percent of e-mails do not replace a letter, the sheer volume of e-mail means that more than 2 billion pieces of mail are diverted electronically each year. And that's just personal correspondence. Between 1999 and 2005, the number of people who opted to pay their bills electronically rather than by mail diverted 3 billion pieces of mail. In the postal world, this replacement of tangible mail with electronic communication is called electronic erosion—and some of this is a good thing. Today we can type a note on our computer in New York and it will be received in New Zealand in nanoseconds. We use e-mail to send documents, music, wills, photographs, spreadsheets, and floor plans, communicate with our banks, send invitations. We no longer have to fill out those irritating forms to receive a return receipt by post, proof that our important letter arrived. The computer does it for us. We can even get a message the moment someone opens our e-mail. In just this one area, e-mail has given us back several days each year.

But it would appear that we are spending that surplus time e-mailing. The average office worker *sends and receives two hundred e-mails a day*—and that figure is rising. Forget about time spent stumbling absentmindedly around the Internet; this habit is destroying our ability to be productive. Information overload is a $650 billion drag on the U.S. economy every year. E-mail has made us a workforce of reactors, racing to keep up with a treadmill pace that is bound for burnout and breakdown and profound anger.

The form's inherent blind spots always catch up with us. According to a survey in the *Journal of Personality and Social Psychology,* we misunderstand the tone of e-mails 50 percent of the time—and for good reason: there is no face on the other end to stop us in midsentence, to indicate that what we are in the process of saying is rude, not comprehended, or cruel. We say what we want, like the CEO who recently belittled the effect of mortgage foreclosures, inadvertently sending the e-mail to someone who had just lost his home. The unlucky call this mistaken judgment. Psychologists call it disinhibition, and its pervasive effect—as can be witnessed every day in nasty comments appended to newspaper articles online, in the aggrieved tone and intent of some blog postings, in e-mail inboxes scorched by flame wars—has turned many parts of the Internet into a nasty place.

It's tempting to simply argue that the Internet attracts aggressive people. But all of us, at some point or other, have behaved poorly over the Internet and via e-mail. There's a reason for these communication hiccups and explosions. According to some neurologists, we learn to interact with the world by mirroring others; not only do we need to see people to understand them most effectively, but our mind learns how to move our limbs and make sense of the world by mirroring the actions of others. There are even neurons in our brain that fire only in response to mirroring the actions of others, and they are intimately connected with the parts of our brain that allow us to move and understand the world. The part of our brain that controls grasping motions shows heightened levels of neural activity when we see someone else pick up a glass of orange juice, as if we were doing it ourselves. According to Marco Iacoboni, professor of psychiatry and behavioral sciences at UCLA, this has bolstered the notion that "our mental processes are shaped by our bodies and by the types of perceptual and motor experiences that are the product of our movement through and interaction with the surrounding world." Consider, then, the ramifications of an era of communication in which we are disembodied as never before. In our new context of e-mail overload, we are working in an environment in which there is nothing to mirror but our own words.

Beating Back the E-mail Tsunami

Who has time to think clearly when under assault by this tsunami of other people's needs? That's what it feels like when you turn on your computer first thing in the morning at the office and find fifty e-mails, the tide of your inbox always rising. One's instinct is to beat it back because e-mail has reoriented time; communication that once took hours, days, minutes, now takes seconds, and the permitted reply time has shrunk as well. Let an e-mail linger for a day, and you risk a rift in a relationship. A 2006 Cisco research paper concluded that failing to respond to a sender can lead to a swift breakdown in trust. Lose an e-mail forever, and you are sitting on an unexploded land mine.

In the past, only a few professions—doctors, plumbers perhaps, emergency service technicians, prime ministers—required this kind of state of being constantly on call. Now almost all of us live this way. Everything must be attended to—and if it isn't, chances are another e-mail will appear in a few hours asking if indeed the first message was received at all.

In the face of this ever-rising onslaught, there appear to be just two choices: keep up at all costs or put up a moat, declare oneself unreachable for the time being—and start all over again. E-mail bankruptcy is the communication subprime mortgage crisis of our era. Ironically, among the first to declare this were the Internet visionaries, such as Lawrence Lessig, founder of the Stanford Law School Center for Internet and Society, who believes that computer code will or can regulate our world as legal code has done in other realms of life. "Dear person who sent me a yet-unanswered e-mail, I apologize, but I am declaring e-mail bankruptcy," he wrote in the summer of 2004. With one quick message, Lessig's correspondents who were waiting for replies became his epistolary creditors, and he pleaded with them just as a bankrupt man does with his lenders. "That's not a promise of a quick response," he continued after five paragraphs. "But it is a promise that I will try." Ironically, his plea for a reprieve generated a "torrent" of new e-mail.

In the beginning, this type of e-mailer—the tech-savvy fellow who sent and received a few hundred e-mails a day—was called a "power user," who took technology and made the most of it. Now every white-collar employee is expected to be one. Not surprisingly, workshops and office coaches will tell you the problem isn't the technology or even the work ethic—it's ourselves. We have bad habits; we reply to all; we waste time treating e-mail as if it were an instant message tool, asking open-ended questions—"How are you doing?"—in the middle of the day. Get it together. You can keep up if you try. But is this really possible when most of us have a water cooler inside our computer surrounded by five thousand people, all talking at once?

In the Western and well-to-do parts of the world, in offices in Dubai and Duluth and Dunkirk, the world's workers are typing themselves into a corner, ever farther out of touch with people beyond their sphere. Walk down a corridor in many companies, and it is eerily silent. You might think it was Christmas morning. In some places, all you hear is the ambient hum of the central air-conditioning unit, the creak of Aeron chairs, the cricketlike click of the mouse, and the faint clatter of keystrokes. But if you lean into cubicles or peer between doorways, you will see hunched, tense figures at their computers frantically trying to keep up with their inboxes. Interrupt them, and you will find their expressions glazed, their eyes dried out and weary. Their keyboard has become a messaging conveyor belt—and there is no break time.

This electronic conversational buzzing has become so loud, it's easy to forget there are people who are not taking part in it. To e-mail one has to be literate, have access to a machine, and be connected. The world's netizen population is approaching 2 billion, but this means only one-third of us are taking part in this enormously useful, endlessly irritating tool. Technology, so often assumed to be the cure for the world's inequalities, has once again simply transplanted them into a new space where English has become the new superlanguage. Africa may be home to 14 percent of the world's population, but it accounts for just 3 percent of the earth's Internet users.

Becoming the Machine

The Corliss steam machine

In 1900, Henry Adams, the grandson of a U.S. president and one of his age's most observant historians, visited the Paris World's Fair and had many of his suspicions of the future confirmed. Standing before a Corliss steam engine like the one pictured above, Adams witnessed the demolition of human narrative, of human scale. Powered by dynamos, huffing away without a single human hand touching its controls, the engine was an enormous testament to the will to power of technology. "Between the dynamo in the gallery of machines outside and the engine house outside," Adams wrote in *The Education of Henry Adams,* "the break in continuity amounted to abysmal fracture." In other words, in this one machine Adams saw how energy that originally would have come from human beings had been replaced by something insentient, a thing that would run itself with no input from human hands except in its creation.

In the twenty-first century, those of us who work in offices have crawled inside the dynamos, the machines driving the system; we're keeping it spinning one electronic message at a time. This symbiotic embrace with the machine is something the early pioneers of the computing age hoped for. J. C. R. Licklidder, a professor of engineering at the Massachusetts Institute of Technology (MIT) and the first director of the Pentagon's Advanced Research Projects Agency, summed up these hopes in a prescient early paper, "Man-Computer Symbiosis": "The hope," he wrote, "is that in not too many years, human brains and computing machines will be coupled . . . tightly, and that the resulting partnership will think as no human brain has ever thought and process data in a way not approached by the information-handling machines we know today."

Fifty years on, that day seems to be here. To read an e-mail, you must be joined to an electronic machine. What does this machine want? Besides following our commands, it is a machine deeply, fundamentally connected to commerce. More often than anything else, it wants us to work. The new on-the-ball employee proves his worth by his speed of response—at work, at night, on the weekends, on vacations, the instant the announcement is made that it is now safe to use approved electronic devices on airplanes.

This ethic of being "always on" extends to the home, where it acquires a consumerist dimension. Web-based e-mail, which is used by more than 1 billion people worldwide, remains free because it allows host companies—such as Yahoo!, Google, and Microsoft—to deliver advertising messages to people refreshing their inbox screen. Every time your screen reloads, a cluster of messages and graphics coalesces in the margins, blinking and beckoning. It frames what you are about to write or read. We are approaching a world in which every letter we write home, every love poem we read, every condolence note, political petition, and letter of apology we type is framed by a penumbra of automobile ads, perfume pitches, entreaties to enter online gambling emporiums.

Faster, Faster

Speed—the god of the twenty-first century—is not a neutral deity, as it turns out. The speed at which we communicate determines what we can do, what we can see, how we perceive, and whether we can adjust our own sense of reality to a larger, more complex frame of reference, one that encompasses the separate needs and points of view of others. Look out a window of a train traveling at full speed, and you will witness this phenomenon at work. The eye constantly darts to the horizon, only to be overwhelmed by a new horizon point, which comes racing forward, followed by another and another. The eye quickly becomes fatigued. The scenery is a blur.

Working at the speed of e-mail is like trying to gain a topographic understanding of our daily landscape from a speeding train—and the consequences for us as workers are profound. Interrupted every thirty seconds or so, our attention spans are fractured into a thousand tiny fragments. The mind is denied the experience of deep flow, when creative ideas flourish and complicated thinking occurs. We become task-oriented, tetchy, terrible at listening as we try to keep up with the computer. The e-mail inbox turns our mental to-do list into a palimpsest—there's always something new and even more urgent erasing what we originally thought was the day's priority. Incoming mail arrives on several different channels—via e-mail, Facebook,

Twitter, instant message—and in this era of backup we're sure that we should keep records of our participation in all these conversations. The result is that at the end of the day we have a few hundred or even a few thousand e-mails still sitting in our inbox.

We're not lazy; the computer is just far better than the human mind at batching and sorting. E-mail travels to and from computers circuitously, starting with our fingers, which type the characters. Our jokes and jabs are eventually translated into 0s and 1s, fired off through cable and phone lines, and reassembled upon the point of arrival, not unlike a car that has been shipped to the United States from Japan in pieces and assembled there once all the parts have arrived at the port and been sent by train to assembly plants, as one technology writer once put it. Computers and e-mail software are designed to know which parts of the chains belong to which; they can wait for a message to arrive fully before delivering it, and they can do so on a scale that is suprahuman. The computer is the ultimate multitasker—it doesn't need to pause to write down reminders to itself on a yellow Post-it note. It doesn't have emotional needs. It doesn't have days when it is depressed. It needn't touch a single thing to feel okay about doing its job.

Look into My Eyes

Don't try this argument out on an Internet visionary. The World Wide Web is often described as the biggest invention aiding human knowledge since the printing press. This may be overblown, since it is impossible to judge at this point—maybe nanotechnology will surpass it, or bioengineering, or battery technology? One thing, however, is clear: the Internet has effected one enormous change in our day-to-day life as it relates to reading, a change so large, but so all encompassing, that we don't notice it—until we step outside.

Since the beginning of time, humans have read by reflected light. This gave reading a sacredness—light, after all, is the first thing God creates in the Bible. In the Koran, "God is the Light of the Heavens and the Earth." Light is a fundamental feature of nearly all founding myths. In Greek mythology, Hyperion, the Titan god of light, is the son of Ourans (Heaven) and Gaia (Earth). In "The Kingdom of the Dead," the gloomiest chapter of Homer's *Odyssey,* his hero washes ashore in a place so wretched that "the Eye of the Sun can never flash his rays through the dark and bring them light." We read to come out of the darkness and into the light.

Before electric light, reading meant sitting by a window or in a room open to sunbeams, or near a candle after dark. Read outside on a park bench in decent weather, and you will realize how natural this feels. The eye is designed for this kind of light, and our chemical response to it regulates our sleep and our moods, gives our days a natural rhythm. Electric light did not change this equation fundamentally. A bank employee might have to read ledgers under a harsher light, a reporter might sit and type a story before a single bulb, but the light they worked by was still reflected, the light glancing down onto the page and bouncing back up into their eyes, at which point the mind can begin to process what's on the page.

The computer screen, however, is an entirely new reading experience. Rather than bouncing down off a surface, light is shot directly into our eyes. It is beamed right into our pupils, and our eyeballs get drier and drier as our blink rate decreases. In the days when computers were used for just word processing, this was not an overwhelming burden. Back then we still read the news and memos and mail in print, by reflected light. But once the Web became so immense it could house large, important sources of information—the home of newspapers, banks, and shopping malls—that were accessed daily, sometimes hourly, the equation changed. And with e-mail, which is checked minute to minute by a great many, that equation exploded. All day long, light is being beamed into our eyes.

Not surprisingly, this accelerating change in how we read has enormous physical and behavioral consequences. Eyesight has deteriorated with the ages, but it has taken a large leap back during the computer age due to the fact that people spend big chunks of their day focusing on a screen that is two feet in front of their faces. There are even nearsightedness epidemics among children. In Singapore, for instance, 80 percent of children are myopic, up from 25 percent just thirty years ago. Close study of books, but also computers and video games, is thought to be to blame.

Our eyes are tired, we get headaches, yet we cannot look away from the screen. E-mail is addictive, it has been shown, in the same way that slot machines are addictive. You press the send/receive button just as a gambler pulls down a slot machine lever, because you know that you will receive a reward (mail/a payout) some of the time. The best way to increase the chance of a reward is to press "Send" a lot. In one study, participants manually checked their e-mail thirty to forty times an hour.

Brave New World—The Lonely Crowd

This shift—the replacement of actual human interaction by a kind of agitated virtual communication, the privileging of the eye above all else—is part of a larger change in our society in which the image has replaced the real thing. We don't stand before art and experience its mysterious aura (as Walter Benjamin discussed in *The Work of Art in the Age of Mechanical Reproduction*) but download its image, buy a poster of it, take home a coffee cup covered with Matisse's goldfish. As Susan Sontag noted in *On Photography,* we cannot travel and be tourists without ferrying home images of the place we have visited—as if the purpose of the trip were the collection of the images, not the being there. Brands exist for this reason. Unable to personally see the tailors or, more typically, the sweatshop labor that goes into making products, we are taught how to identify a brand and its logo with particular traits we prize: Audi's linked circles are the mark of engineering precision; Starbucks' cup goddess is proof that a multinational coffee chain has bohemian roots.

When we are pummeled by ads, awash in representations of the world, is it any surprise that the real-world commons—a shared space in which people of all sorts can meet and interact—has been shunted aside for its electronic simulacra? Instead of driving down the road to our local bookstore, where we might actually talk to someone, we buy a book over Amazon.com or Barnesandnoble.com; rather than go into the bank, we check our balance from home; rather than buy the newspaper from a paperboy who comes to collect the monthly bill, we read it online, for free. These are all conveniences, significant ones for the busy, for people who live in remote locations, or for people for whom face-to-face conversation is inordinately stressful, but the upshot is that we spend less time dealing face-to-face with other human beings and more time before a machine.

Thirty years ago, in *The Society of the Spectacle,* the French philosopher Guy Debord predicted we would be spending more time apart. "The reigning economic system is founded on isolation," he wrote. "At the same time it is a circular process designed to produce isolation. Isolation underpins technology, and technology isolates in its turn; all goods proposed by the spectacular system, from cars to televisions, also serve as weapons for that system, as it strives to reinforce the isolation of 'the lonely crowd.'" To this list of machines we can now also add the Internet and e-mail.

Ironically, tools meant to connect us are enabling us to spend even more time apart.

The most glaring discovery of the Stanford University study mentioned earlier was not that people burned up two hours a day on the Internet but that those two hours came out of time they would normally spend with family and friends. Once that withdrawal has begun and technology has been identified as a way to connect, it's a hard cycle to break. We blog, broadcast our vacations on YouTube, obsessively update the newsfeeds of our Facebook pages—"Today, Brian is feeling happy"—as if an experience, an emotion, a task completed hasn't actually happened unless it has been recorded and shared with others. E-mail is the biggest, broadest highway on which this outward projection occurs. Why write a postcard about your trip to France to one friend when you can simply forward and copy the message to all your friends? Why tell a coworker you have performed an arduous piece of labor when you can cc several others and make sure they know it, too?

In the twenty-first century, writing and "publishing" have become easier than ever—and reading, due to the amount of material available to read and the rate at which we are communicating, has become harder than ever. This wouldn't be quite so untenable an environment if we were actually seeing each other face-to-face. But the drop in face-to-face contact has taken this epistemological fracture and given it an emotional dimension. We have all the tools in the world, yet we've never felt more alone. By depriving ourselves of facial expressions and the tangible frisson of physical contact, we are facing a terrible loss of meaning in individual life. The difference between a smiley face and an actual smile is too large to calculate. Nothing—especially "lol"—can quite convey the sound of a friend's laughter.

Talking Back

On a small scale, perhaps this model of frenzied communication would work. Think of a house in which six roommates share everything and anything and the closeness this fosters. But ironically, due to the networked, interlinked nature of the Internet and the way it grows virally, exponentially, this constant chatter is utterly unsustainable. The creeping tyranny of e-mail is a symptom of how out of control the situation has become and it is only going to get worse as more and more people around the world get broadband and e-mail accounts, and multinational companies, which rely on workers in different parts of the globe staying in touch, expand and put down even larger global footprints in the real world, not to mention in the cloud of machines connected to the Internet. We are at the beginning, not the end, of this problem.

The tyranny of e-mail has also entered a feedback cycle that makes it ever harder to reflect on how bad the situation has become. Spending our days communicating through this medium, which by virtue of its sheer volume forces us to talk in short bursts, we are slowly eroding our ability to explain—in a careful, complex way—why it is so wrong for us and to complain, resist, or redesign our workdays so that they are manageable. This book is an attempt to slow things down for a moment so we can look at the enormous shift in time and space e-mail has effected, how e-mail has changed our lives, our culture and workplace, our psychological well-being. No one can predict the future of a technology, and this book is certainly not going to try, but it is essential, especially when that technology has become as prevalent and pervasive as e-mail, to examine its effects and assumptions and make an attempt to understand it in a broader context.

We are evidently remaking our environment, so it's fair to ask: What does this new world look like? What are its roots? How does the technology upon which it runs affect what we can say or how we say it? Should we have a correspondence list in the thousands? Does this way of living seem natural or even sustainable? Surrounded by the plastics, polystyrenes, and chemicals of the modern workplace, our bodies have an instinctual memory of something more natural. This metaphysical nostalgia, which Alan Weisman beautifully describes in *The World Without Us,* is a source of profound anxiety, and not the kind that can be medicated or wrested into submission. Speed cannot mask this anxiety, either; it only destroys our ability to reconnect with something actual.

Ever since humans emerged from Plato's cave, we have tried to communicate with each other. Sounds turned into pictures, which turned into phonetics, which were eventually written down and codified, printed on clay, then parchment, then on paper. Mail has existed since at least the ninth century in Persia. The printing press allowed a person to address a multitude without being there to say it to them (or copy it by hand). It took hundreds of years, however, for books to become widely accessible. And it took yet more time for those books and newspapers and letters to be shipped from one city or continent to the next. And then societies had to help their citizens become literate for these publications to be read in large numbers.

In the nineteenth and twentieth centuries, however, we leapt from the speed of transport to the speed of electricity. The telegram allowed people to address each other one to one, within a day, at a price so cheap it eclipsed that of the long-distance phone call. Twenty million telegrams were sent in 1929 alone, this when the world's population was 1.5 billion. Today, the world is home to 6 billion people and *roughly 600 million e-mails are sent every ten minutes*. Stop for a moment to imagine the ramifications of this exponential increase in communication, and the necessity for a pause cries out like an air-raid siren.

Previous generations, however giddy they became about the best technology, did stop and think—if briefly. Samuel F. B. Morse sent the first telegram to go through in the United States, from Washington to Baltimore, in May 1844, with the message WHAT HATH GOD WROUGHT. By contrast, the first e-mail ever sent using the @ symbol was mailed from one supercomputer to the next in all caps, and according to Ray Tomlinson, the man who sent it, the message contained just a random series of letters and numbers. In other words: gibberish. He just wanted to see if it would arrive and so didn't bother to type anything providential.

It's about time we asked ourselves a more articulate question: What have we wrought? To answer it, we're going to have to go back a long way, not just to the dawn of the first information age, when people first began communicating at the speed of electricity, but even further, to when people the world over were just beginning to get mail, and see what happened when the dream of obliterating distance started to become reality.

What follows in the next three chapters is a brief, selective history of how we went from reed stylus to silicon computer chip. At each step of the way, the new manner in which words moved over space introduced a new experience of reality, one that gradually built up to an experience of overload. The democratization of words through education and mail unleashed a blizzard of letters; no longer were they written and read only by a few. The creation of the telegram linked the world by a wire, and the people at home reading newspapers, expanded by telegraphic reports, suddenly had to—like operators for Western Union—tell signal from noise in an entirely different news environment. The creation of the Internet and the PC simultaneously made every inbox a telegram portal of a late-twentieth-century sort; it finally brought about the dream of obliterating distance.

All these developments have brought us to where we are today; in each period governments and crooks have attempted to stay one step ahead of the curve to exploit the increased amount of human traffic going over roads, wires, or T1 cables. Each communication breakthrough has encouraged individuality while expanding the notion of the commons beyond the tangible or nearby.

But for many of us the creation of the Internet has done one thing none of these leaps forward in communication history could: it has tied us irrevocably, perhaps fatally, to a machine and its superhuman capability. If we are to understand our predicament today, we must reckon with the changes that working at this machine has wrought and examine whether there is a way we can slow down, so we can make the best use of it while retaining a foothold in the real-world commons. Otherwise, we will have bridged the darkness only to introduce ourselves into one of another, more relentless kind.

1

WORDS IN MOTION

I watched a letter that I had written start off on its journey in a howling snowstorm, high in the mountains of Finnish Lapland. The postman was a gnarled little Lapp and his means of transport a flat-bottomed sleigh, drawn by a reindeer. . . . After three or four hours and perhaps a spill or two in the snow, it would travel five hours by bus and then a day and a night by train to the Finnish capital. From there it would go by ship, steaming in the channel cut by an ice-breaker through the frozen sea to Sweden. Swedish postmen would convey it across their country and put it on an ocean liner. On arrival in New York, the United States Post Office would take charge, and finally an American rural postman would deliver my letter on his rounds in a small Midwestern village. I had paid the equivalent of fourpence for all this service. What is more, I had paid the Finnish Government alone, not the Swedes or Americans.

—Laurin Zilliacus, *From Pillar to Post: The Troubled History of the Mail*

The success of mail in modern society is, to this day, something of a marvel. By our simply dropping a letter into an iron box, an object we have held and inscribed with words that only we can form may travel around the globe in a matter of days. On such a journey mail has been carried by foot and by horse, by chariot and by pigeon; it's journeyed by balloon and by bicycle, by train, truck, steamboat, pneumatic tubes, airplane, and even missiles. If a thing can move, it has probably carried mail. Bringing this service to people without a title took thousands of years. Roads had to be carved into fields and blasted through mountain passes; durable, cheap writing utensils had to be invented and men and women organized into labor forces. To this day, the U.S. Postal Service is the second-largest employer of civilian labor in America, with nearly *800,000 employees* (compared with 181,000 in the UK, 160,000 in Germany, and 100,000 in France).

In ancient civilizations, mail was haphazard for most people—a luxury, a stab in the dark. So the act of sending and receiving a letter was a momentous occasion. Without the highly efficient systems we possess today, letters were delivered informally, through friends or acquaintances planning to travel to a certain location. Along the way the messenger might be robbed, injured, or killed; the absence of a reply was so common that many ancient letters contained complaints about the failure of the recipient to respond to a missive. It would be hard to blame the messenger, though.

There was very little letter technology. Although clay envelopes dating back to several thousand years B.C. have turned up in Turkey, the paper envelope is a recent invention. Until the late 1800s, schoolchildren in America learned to fold a letter so that it didn't need one. The stamp, which came into being in India and England in the 1840s, is new, too, along with paper (which appeared in 3500 B.C.), to say nothing of pens (an invention of Egypt in 3000 B.C.) and zip codes (first used in the United States in 1963). Prior to all these developments, people addressed letters on the back of the missive itself or right on top of the text. Seals were important for this reason: they were a stamp of authenticity.

The history of mail is a tale of how, with the invention of postal systems and the democratization of their use, words began to knit more than just nations together. Words written by hand, then carried by the saddlebags of travelers, kept friendships alive and gave shape and texture to the daily experiences—and the thoughts—of people who wanted to communicate but were not within speaking distance of one another. In arcing across that gap, letters and mail helped create (and remind us of) another gap—the one between the inside and outside world. "It's separation that weaves the intrinsic world," Hélène Cixous writes of letters. "A fine, tender separation . . . like an amniotic membrane that lets the sound of blood pass through."

Mail has been the world's most important artery for transmitting our pulse across that separation. As words and then language were democratized and mail extended to larger parts of the world's population, that sound has become louder, syncopated, cacophonous. Governments and businesses have listened in; tricksters and thieves have posed as lovers and bearers of good news. Our current age is not the first in which people struggled to keep their inbox tidy. And speed, that perpetual, beckoning messenger, threatens to obliterate the very thing—distance—that made us want to write to begin with. But we're getting ahead of ourselves. First, the mail just had to get there.

The Pillar of Empires

In the beginning, mail was a tool of a few—of governments and militaries and kings. It's not hard to fathom its usefulness.

A well-organized postal service knit enormous stretches of land together; it announced news of battles lost and won, collected intelligence, and delivered the occasional expression of courtly love. Not surprisingly, all the major early empires had some system of carrying letters from one place to the next: the Aztecs, the Incas, the Chinese, the Assyrians, the Romans, the Mauryans. In most cases only government officials could use the service. This was not an enormous deprivation, as most citizens could not read or write. In very important circumstances, they could have letters written for and read to them.

In the sixth century B.C., the Persian Empire under Cyrus the Great boasted a well-organized relay service that could carry mail at a rate of up to a hundred miles a day. The man carrying the message would ride from one post to the next, where he would trade his tired horse for a fresh one, rest, then continue upon his journey. Herodotus sang this system's praises when he wrote in the histories a comment now inscribed on the Farley Post Office Building in New York City: "Neither snow, nor rain, nor heat, nor gloom of night stays these couriers from the swift completion of their appointed rounds."

The Persians' mail service, like so many, was not a benign network, however. It was a tool for making war and controlling an expanding population, upon whom the postal carriers spied and reported back. All the emperors did it. Kublai Khan had more than ten thousand postal stations and some fifty thousand horses at his disposal—and the people who lived near the postal stations learned to fear the carriers. Caliph Abu Jafar Mansur, who ruled the Arabian Empire in the eighth century, expressed mail's military importance most bluntly: "My throne rests on four pillars and my rule on four persons: a blameless cadi [chief justice], an energetic chief of police, an honest minister of finance, and a faithful postmaster, who gives me true information about everything." In 860, the Islamic caliphate boasted 930 post stations.

News of military victories traveled via these early official postal systems, as did instructions for slaughter. As Laurin Zilliacus reminds us, the Book of Esther describes "the use of posts to order the slaughter of the Jews throughout Persian-ruled territory, and then the swift sending of the counter-order that saved them and turned the tables on their persecutors."

"And he wrote in the name of the king Ahasueres and sealed it with the king's ring, and sent letters by posts on horseback," goes the verse, "riding on swift steeds that were used in the king's service, bred of the stud." All of the books of the New Testament, save the Gospels, are written in the form of letters.

It was not always horses doing the carrying, however. The Arabs later pioneered the use of pigeons. The Greek city-states reserved some of their loftiest poetry for the mind-boggling feats of their athlete-runners. They were known as Hemerodromes and were often called into service when a very important message was to be delivered. Philonides, the courier and surveyor for Alexander the Great, once ran from Sicyon to Elis—148 miles—in a day. Wealthy families had stables of runners, and since nearly all the work was taken care of by their slaves, citizens of leisure had little to do but share their thoughts and gossip in letters to one another. Not surprisingly, a huge volume of correspondence was produced during the height of Greek civilization.

Augustus Caesar established one of the most impressive ancient postal services. It relied upon the Roman Empire's superior road network, with stopping-off points—or post houses—where couriers could rest and trade horses. It is for this reason that the word "post" comes from the Latin *positus,* meaning "fixed," or "placed." Mail traveled by horse and chariot, and the postmen of the era wore feathers in their caps, signifying speed. The service was eventually expanded to the general public—those who could write and afford it. When the Roman Empire fell, the network collapsed, and organized communication throughout western Europe disappeared with it.

Filling the Gap Where Governments Leave Off

Guilds, trading companies, feudal lords, and marauding armies maintained private messaging systems after the fall of the Roman Empire, but only the Catholic Church possessed anything approaching the organization of the service Caesar had run from Rome. During the medieval period apostolic and pastoral letters "circulated doctrinal rulings, decisions of Episcopal synods, temporal and political matters," as University of California professor Charles Bazerman has written. The growing orders of monks also kept in touch through lay brothers traveling from one monastery to the next, trips that could take as long as several months, carrying scrolls called *rotulae,* an early, low-tech version of a Listserv. A scroll would leave a central monastery with simply, say, a list of names of brothers or benefactors who had died and ought to be remembered. At each monastery an addition would be made to the scroll, the local abbot acknowledging receipt of the message and perhaps adding comments or further news. The additions were attached by "thin intertwining parchment strips, so that the lengthening scroll continued to be a very long single sheet," as Zilliacus describes. One *rotula* dating from 1122 is 28 feet long and 10 inches wide, covered with writing on both sides. It contained one piece of news, the death of Abbot Saint Vital; all 206 entries that followed paid tribute, some in prayers, some in poetry.

In Europe, a single family kept mail going outside the realm of government for several centuries. The princely Thurn und Taxis clan carried the mail from Rome to Brussels and beyond, beginning with Ruggiano de Tassis, who began a postal service in Italy. In the early 1500s, they established a postal service based in Brussels and reaching to Rome, Naples, Spain, Germany, and France by courier. The service lasted until the eighteenth century, when it was purchased by an heir to the Spanish throne.

> *But the story of the post is not concerned only with past events. It is a continuing saga of man's attempts to shrink the world by improving the communication of the written word.*
>
> —Dmitry Kandaouroff, *Collecting Postal History: Postmarks, Cards and Covers*

Bridging the Darkness

It took another three centuries for mail to resemble anything we would recognize today. Most people were illiterate, and paper was enormously expensive. For a woman to mail a letter from the American colonies to England in 1650, for example, she would have had to be exceptional indeed. In *From Pillar to Post,* Zilliacus imagines such a letter's journey. The woman would have had to pick out a piece of rag paper, inscribe her message, then fold it four or five times, bind it in silk and seal it with wax, and perhaps draw a sign of the cross on it to signal that this letter traveled under the sign of Providence. Then there was the issue of the recipient's address, which required a bit of space: "To my most noble brother, Mr. John Miles Breton," wrote one woman, "at Ye barber shoppe which lieth in the land hard against Ye taverne of Ye Great Square in shadow of Ye Towne Hall in Stockholm, these."

After she took it to her local tavern—America's first postman, Richard Fairbanks, pulled pints as his day job—the letter would have to jump aboard a trading vessel and essentially bribe its way across the seas and every step of the way. It was expensive and very likely bound for failure. "Letters were treasured," writes Frances Austin in "Letter Writing in a Cornish Community in the 1790s," "read to neighbors and handed round to friends. Items of news were passed on, often almost verbatim." In one case, Austin discovered that a woman had copied a letter from her brother and mailed copies to each of her siblings.

Imagine, then, living at this time. If you emigrated to a new country and left relatives behind, they would in all likelihood be lost to you forever. At the close of day, eating by candlelight, going to sleep in the obsidian darkness beneath a sky punctured by a blizzard of stars and unmarred by the electric pulse of distant cities or the blinking of far-off satellites, you would be alone, save for those around you. A knock at the door could spell danger or bad news. If the visit brought a letter from far away, it would probably feel like a small miracle. "The letters delivered in the countryside have marvelously multiplied," wrote Richard Jeffries in *The Life of the Fields* in 1884, "but still the country people do not treat letters offhand. The arrival of a letter or two is still an event; it is read twice or three times, put in a pocket and looked at again."

The Dawn of the Golden Age of Letter Writing

Between the dawn of the printing press and the end of the early to middle nineteenth century, societies arose in England,

France, and Germany that broke down the barriers that had previously kept many people in the dark concerning the written word. Books became more available and began crossing national boundaries. By 1873, more than 129 million book packets traveled through the post every year in Great Britain.

The printing press and changes in the epistemological makeup of religions forged the way for a new reading public. In many Christian religions, only pastors and priests were trusted with the word of God. Besides, books were scarce; monks used to teach from texts chained to the lectern. In Sweden in the eighteenth century, though, the Lutheran Church issued an injunction that everyone must be able to read the word of God, and a massive literacy campaign was launched. Within a hundred years the nation boasted a 100 percent literacy rate. The decline of illiteracy in England was equally sharp but took much longer. Between 1500 and 1900, literacy rates rose from 10 percent to 95 percent for men and from less than 5 percent to 95 percent for women. In the early American colonies, where religious injunction required that believers be able to read the Bible themselves, men had a 100 percent literacy rate.

Mandatory schooling helped democratize words, too. In 1790, "Pennsylvania made provision in its constitution to provide free education to the poor, an effort endorsed by a number of cities and states in the first half of the nineteenth century," according to Naomi S. Baron in *From Alphabet to Email.* In 1827, Massachusetts underwrote common schools through taxation, and in 1852 New York was the first state to require statewide compulsory education. The earliest English public schools were chartered to educate the poor, but it wasn't until the middle of the eighteenth century that religious organizers began providing systematic day school and Sunday school education. Parliament started funding schools in 1833, while also restricting child labor, and by 1880, all of England and Wales "were required to establish minimum education standards."

Letter-writing manuals sprang up to allow the "middling sort" to "pursue their claims to social refinement and upward mobility," as one scholar put it. One of the most popular, *Letters Written to and for Particular Friends, on the Most Important Occasions,* was published by Samuel Richardson in 1741 and portrayed letter writing as "suitable for all occasions in life and for all people in society." At the same time, penmanship manuals, spelling books, grammar books, and dictionaries took off. Between 1750 and 1800, nearly four hundred such works were published in the United States alone. Guidebooks to letter writing multiplied, in the process transforming the use of the English language. The oral quality of early letters returned and was encouraged: "When you write to a friend," wrote W. H. Dilworth in *The Complete Letter-Writer,* "your letter should be a picture of your heart." "When you sit down to write a letter," another advised, "remember that this sort of writing should be like conversation." Dilworth again: "Your language should be so natural . . . the thoughts may seem to have been conceived in the very words . . . and your sentiments to have sprung up naturally like lilies of the field."

In a new development and with increasing regularity, these appeals were directed at women. Until this point, it was generally assumed that only men wrote letters. But from the mid–eighteenth century, the gender division of letter writing began to be questioned publicly. *The Ladies Complete Letter Writer* appeared in London in 1763 and was imported to America. "Your sex," wrote one adviser, "much excels our own, in the ease and graces of epistolary correspondence."

Letter writing also became an important part of childhood instruction, in both England and America. Benjamin Franklin believed young boys should learn by putting pen to paper to send a message. "The Boys should be put on Writing Letters to each other on any common Occurences, and on various Subjects, imaginary Business, & c. containing little Stories," he wrote in *Idea of the English School* in 1751. "Accounts of their late Reading, what Parts of Authors please them, and why. . . . In these they should be taught to express themselves clearly, concisely, and naturally, without affected Words or high-flown Phrases."

Franklin extended such advice to writers of business letters. In April 1777, he replied at length to a gentleman who had written offering some assistance to the American colonies in the coming War of Independence: "Whoever writes to a Stranger should observe 3 Points; 1. That what he proposes be practicable. 2. His Propositions should be made in explicit Terms so as to be easily understood. 3. What he desires should be in itself reasonable. Hereby he will give a favourable Impression of his Understanding, and create a Desire of further Acquaintance."

Inventing the Modern Post Office

For letter writing to really explode, however, it needed to be affordable to the masses. Charles I was the first monarch to extend mail service to his subjects in 1635, largely because he needed money, and even then it was too expensive for most people to use it. In 1680, William Dockwra, an ex-merchant in the African slave trade, set up a penny post in London. For the first time, anyone could mail a letter anywhere in the city for a penny—the equivalent of roughly half a pound, or $1, today—which was a boon to business but not much help to people whose relatives and friends lived a hundred miles away.

The government absorbed the service and then later squashed an attempt to establish a half-penny post in London. Sending letters was still very expensive, since rates were charged by distance and letters had to travel by armed coach. Members of the House of Commons and the House of Lords, however, could use the service for free.

Most of the features of modern mail come from the suggestions of a retired schoolteacher named Rowland Hill, who in 1837 published *Post Office Reform: Its Importance and Practicality.* Hill argued that mail should cost a penny, wherever it went; he believed the postage should be paid in advance; and he wisely suggested that envelopes be used. By 1840, every one of these suggestions had been taken to heart by the British postal service, the Royal Mail, and letter writing had exploded. Between 1839 and 1853, English letter volume shot up sevenfold, from 75.9 million per year to 410.8 million. By 1873, the Royal Mail was handling 1 billion pieces of mail, employing 42,000 men and women, and boasting more than 12,000 post offices.

The Royal Mail quickly went global. As early as 1860, mail traveled once a month via the Suez Canal between Great Britain and the Australian colonies, including Tasmania and New Zealand. It was dispatched from Southampton on the twelfth of each month. Postage for these international shipments had to be paid in advance: it was thirty-three cents per half ounce for a letter. Newspapers were sent for just four cents each.

Most impressively, the Royal Mail made money doing this. In 1860, a report showed that it had earned $6.5 million for the government. By 1873, the sum was creeping up to $8 million. The service became the model for mail systems the world around, its facility paving the way for a heyday of print. People wrote and read more and thereby began to develop a sense of their own ideas. The events of their lives mattered because they were being recorded. They wanted to be heard. Letters pages, like those Richard Steele had inaugurated in *The Tatler* and *The Spectator* in the previous century, filled with an array of new voices.

British citizens also seemed to enjoy testing the limits of this new civic invention. A report in 1874 revealed that among the curios to turn up in the dead-letter office were a horned frog (alive), a still squirming stage beetle, white mice, snails, an owl, a kingfisher, a rat, carving knives, a fork, a gun, and cartridges. One dead letter turned out to have more than £2,000 in banknotes in it; another, which arrived opened, was stuffed with Turkish currency. Thought to be old lottery forms, it was given to children of postal officers to play with.

The Business of Sending Mail in America

Tasked with covering enormous distances in a countryside prone to violence, where gun ownership was a constitutionally defended right, and a strong federal government was deeply distrusted, the U.S. mail faced far greater problems than the British. Early mail routes were marauded; carriers, who had to ferry all letters COD, many times couldn't collect or simply pocketed the payments; bootleg companies covered the same terrain, often at cheaper rates. The postal service was also seen as a sinecure, a system of patronage. Far more letters than should have been were franked or sent for free, depriving the fledgling network of much-needed revenue. As postmaster general, Benjamin Franklin appointed his older brother postmaster of Boston; when James Franklin stepped down from the $1,000-a-year post, Benjamin Franklin appointed his brother-in-law as his successor.

And mail remained, for a long time, hideously expensive. In the early 1800s, it cost twenty-five cents to send a single sheet of paper more than four hundred miles. In a country where the average wage was a dollar a day this was beyond the reach of most Americans, and, not surprisingly, mail volume remained very low. In 1815, the entire United States was serviced by just three thousand post offices, doing a little more than $1 million of business. Two decades later, mail began traveling by rail and then by private stagecoach companies, operating under contracts that were so lucrative that if a company failed to get them renewed it immediately retired from the business. But the two coasts remained separate. To get to California, mail had to travel around Cape Horn, after which coaches and trains took it to the Oregon Territory.

One of the most famous chapters of U.S. mail history involved the shrinkage of delivery times across the country to a matter of days—a vast improvement, considering that in 1845 it took President James Polk six months to get a message to California, and a necessary one financially once gold was discovered in the state in 1848 and San Francisco's population rocketed from 500 to 150,000 in just three decades. News of discoveries had to travel somehow. The problem was serious enough that in 1855 Congress even allocated $35,000 to exploring the use of camels to haul mail from Texas to California.

As has happened throughout the story of communication in America, government turned to private industry to solve the problem. Postmaster General Aaron Brown later awarded a $600,000 contract to the stagecoach entrepreneur John Butterfield to carry mail end to end if he could do it in twenty-five days. By 1858, after spending $1 million to set up a network of two hundred relay stations, two thousand horses and mules, and more than twelve hundred employees, Butterfield had done it. There was even a manual of employee behavior, in which lay this recommendation: "18.—INDIANS. A good look-out should be kept for Indians. No intercourse should be had with them, but let them alone; by no means annoy or wrong them."

Coaches left from Saint Louis and arrived in San Francisco, carting mostly mail but passengers, too, in quarters cramped enough to make airline "coach" seats seem a luxury—especially as coach travelers paid $200 in 1860, the equivalent of $4,500 today, for a one-way fare from Saint Louis to San Francisco. Raphael Pumpelly, who rode the line west to Tucson, remembered, "As the occupants of the front and middle seats faced each other, it was necessary for these six people to interlock their knees; and there being room inside for only ten of the twelve legs, each side of the coach was graced by a foot, now dangling near the wheel, now trying in vain to find a place of support."

But for those with money to spend, a need to travel, and a desire for adventure, it was worth the discomfort. The travel also facilitated the exploration and imagining of the American landscape. In *Roughing It,* Mark Twain described traveling around the western part of the United States on several coaches just like the one Pumpelly packed himself into, though not of the Butterfield line. "We three were the only passengers, this trip," he wrote. "We sat on the back seat, inside. About all the rest of the coach was full of mail bags—for we had three days' delayed mails with us. Almost touching our knees, a perpendicular wall of mail matter rose up to the roof. There was a great pile of it strapped on top of the stage, and both the fore and hind boots were full. We had twenty-seven hundred pounds of it aboard."

It says a lot about America's idea of itself that Butterfield's sensible, pioneering line—which never suffered an Indian attack in the two and a half years of its operation and never missed its twenty-five-day deadline—was eclipsed by the even shorter-lived Pony Express. Speed and brutality trumps efficiency in the imagination. This colorful solution was proposed by California senator William Gwin, a pro-slavery southerner who once participated in a duel (neither party suffered a gunshot, but a donkey was killed) and later traveled to France in 1864 in an attempt to interest Napoleon III in settling American slave owners in Sonora, Mexico. Gwin was a strong proponent of westward expansion and had carried through the U.S. Senate a bill appropriating money for steamers traveling between California, China, and Japan. He was also an early proponent of purchasing Alaska from the tsar. His notion of mail would be similarly spectacular.

In January 1860, Gwin met with the Missouri freighter William H. Russell to discuss establishing a ten-day relay service to California. Gwin was enthusiastic enough to encourage them to go ahead; they had sixty days to do the job. Ads for riders went out in March of that year. They recruited 80 "skinny" young fellows, whose weight was not to exceed 125 pounds, and 400 tough characters to staff 190 relay stations over a 1,900-mile route from Saint Joseph, Missouri, to Sacramento, California. It was dangerous work, for which the young men were paid $50 per month plus room and board. In the beginning, the riders rode with bow knives, revolvers, and at least one rifle, eventually thinning down to just a pistol.

It was also expensive: $5 per half ounce. The saddlebags contained letters, some relaying news of gold discoveries in California, and condensed versions of eastern newspapers. In just 18 months of business, the Pony Express transported 30,000 pieces of mail a total of 650,000 miles. Its fastest delivery set a record in carrying President Abraham Lincoln's inaugural address from Saint Joseph to Sacramento in seven days and seventeen hours.

If only Lincoln had had Obama's BlackBerry! At just fifteen years of age, William F. Cody was the youngest pony rider to carry the mail. On his second year on the job, he rode up to a station and, discovering that his relief rider had been killed, rode on with a new horse, covering some 384 miles without rest.

Once the telegram reached California, however, the legendary service soon became cost-inefficient. The last Pony Express ran in October 1861, just eighteen months after its beginning. Its accomplishments were eulogized in the *Sacramento Bee:* "Farewell Pony . . . Farewell and forever, thou staunch, wilderness-overcoming, swift-footed messenger. . . . Thou wert the pioneer of the continent in the rapid transmission of intelligence between its people, and have dragged in your train the lightning itself, which in good time, will be followed by steam communication by rail. Rest upon your honors; be satisfied with them, your destiny has been fulfilled—a new and higher power has superseded you."

Making a Nation with Words

Around the time the telegram began to take flight, postal rates plummeted. The Postal Acts of 1845 and 1851 reduced the cost of a letter to a flat three cents to anywhere in the United States. The rate wasn't raised again until 1958, when it climbed to four cents. The effect on mail volume was overwhelming. Residential post office boxes, a hallmark of America, began going up in 1858 and soon became ubiquitous. In 1840, the average American sent three letters a year; by 1900 that figure was sixty-nine letters per annum and the total volume of letters outnumbered telegrams fifty to one. By 1950, the mail was almost out of control; in 1960, the U.S. Post Office was handling 63 billion pieces of mail—the equivalent of 350 pieces per year for every man, woman, and child in America.

From the Outpost to the Common

In the rapidly expanding country, the post office became one of the most important public commons. "Most early postmasters were storekeepers," wrote an ex–postmaster general, Arthur Summerfield, in his book *U.S. Post Office.* "Their places of business were the community centers. They knew everyone in town and the surrounding countryside. They were respected. They knew something of elementary account-keeping. Having property, they were responsible and could be bonded."

Mail carriers became people of note, and from a very early time some of them were women. When Franklin's brother John died on the job in Boston, Mrs. John Franklin became not just the first woman postmaster, but also the first woman to hold public office in America. Women postmasters followed in Baltimore (1775) and Charleston, Maryland (1786). The first female carrier went to work in North Carolina in 1794. By 1893, there were 6,335 postmistresses, some of whom juggled several jobs at once, like this woman, described in an early report about mail carriers around the United States:

> Mrs. Clara Carter of West Ellsworth, Maine drives the mail coach from that place to Ellsworth, seven miles away. . . .
>
> This energetic woman rises early in the morning, does the cooking for five in the family, starts at seven for the city with the mail and numerous errands that are given to her without memoranda. She returns at

> noon, gets dinner, goes to the blueberry fields and picks ten quarters of berries or more in the afternoon, and in the cool of evening does the family washing and ironing and other household tasks. This amount of work she performs six days in the week, varying the routine in the afternoon, out of berry season, by sewing for the family. She finds time, too, to play on the parlor organ an hour or more in the evening, or to entertain visitors.

Prior to rural phone service, the postman "would carry news of forest fires, of accident, or an outbreak of illness on a farm, to the nearest communication center. Unofficially he (or she) became the bearer of local news, or gossip if you wish." People who read the news wrote to one another about what they learned, especially emigrants in America; it was a way to connect the past with the present over here with what was once home. "I see in the newspaper that they have had some trouble between Sweden and Norway," wrote Olaf Larsson from Kellogg, Idaho, in a 1905 letter.

The mail was also a highly effective tool at keeping new emigrants in touch with one another. Here's a letter Pet Stred sent to his brother in Sweden from Bay Horse, Idaho, in 1897:

> I am well and work and grind away a little every day but I have been sick, not so that I was bedridden, but I was still very ill a few days ago but am now completely healthy again. I have worked at various jobs this summer. For a while I worked on a road that was being built. One month I worked for a farmer and now I work at a smelter where they smelt ore that is taken out of the mines. That is hard work and takes

> real Swedish strength to bear with it. The work is very tough and it is hot like a certain *place* [meaning Hell; underlined in original]. I do not know how long I will stay here, when I get tired of it [I] will have to try something else. Those who are young and inexperienced should try everything.

As in England, some people enjoyed seeing just how far the mail could go. In 1903, one man set up a Nonsense Correspondence Club and began sending unusual items through the mail. "I owed a friend a dollar," the man wrote. "I mailed him a silver dollar with a two cent stamp stuck on one side and the address on the other. He received it." His next prank, however, created more havoc. When dining in Key West, he filched some croquettes off the dinner table, wrapped them in tinfoil, and mailed them to Philadelphia, labeled NATURAL HISTORY SPECIMENS. The packet burst in the mail, and, worried that it had scattered someone's remains, the Philadelphia post office sent the remains to the morgue and placed them on ice. It then mailed him a letter saying he would be responsible for the cost of this treatment.

The Fastwriter

As the range of communication options proliferated and messages traveled ever faster, two inventions made them go swifter yet. The first actually made it easier to write, once people learned how to use the darn thing. Although the earliest model dates back to 1714, the typewriter was finally perfected in 1868 by a newspaperman, printer, and politician named Christopher Latham Sholes. He tried to sell the rights to manufacture the machine to Western Union, which turned him down, eventually settling with the Remington Arms Company, which made farm machinery, sewing machines, and, most famously, guns.

Remington took the typewriter to market in 1873, and it raised a stir at the Centennial Exhibition in Philadelphia in 1876, while the Hammond typewriter—the first single-element machine, which featured a curved keyboard—stole the show at the Paris Exposition of 1889.

An 1874 Remington

Mark Twain purchased one of Remington's earliest models in 1874 for $125 and became the first author to submit a manuscript typed on one. In a letter to his brother, Twain described the way the machine established an element of speed in writing that had not yet been there before—even though one could, as yet, type only in capital letters:

> I AM TRYING TO GET THE HAND OF THIS
> NEW FANGLED WRITING MACHINE. BUT I

> AM NOT MAKING A SHINING SUCCESS OF IT. . . . I PERCEIVE I SHALL SOON & EASILY ACQUIRE A FINE FACILITY IN ITS USE. . . . I BELIEVE [THE MACHINE] WILL PRINT FASTER THAN I CAN WRITE.

Aside from Twain, most of the typewriter's early users were not authors, however, but rather stenographers and typists, whose numbers in America shot up from 154 in 1870 to 112,364 in 1900. Many of them were women. "Some of the more enterprising of the girls secure an office in a big building, where lawyers are numerous, put out a sign, and find employment all day long," wrote a reporter in *The New York Times*. "The regularly-employed girls get $10 and $12 a week, but the owners of the machines manage in some cases to earn so much as $20 and $25."

And they typed fast. To be a member of the New York State Stenographers' Association, for example, one had to be able to take dictation at 150 words per minute for five consecutive minutes. The first woman ever hired by New York City's Health Department was Miss Martha N. Manning, a typist. Typewriting contests began to be held. One of the earliest was won in New York City by F. E. McGurrin of Salt Lake City, who then closed the contest with an encore act of typewriting 101: words a minute while blindfolded. In 1889, a woman named Miss M. E. Orr "made her fingers fly over the keys for a minute and 139 correctly printed words was the result."

Henry James briefly made dictation a craze when, to overcome a writer's block, he hired an amanuensis to take dictation. He found that he wrote more, his sentences grew longer, and he could work only to the rhythmic click of the Remington. The shadow of this method was large enough that when William

Dean Howells was interviewed by *The New York Times* in 1882, he was asked if he, too, wrote by dictation. "I do not dictate," he said, "but use a little Hall typewriter. I use it entirely if I have a clear block of stuff before me; if I have to come down to close quarters and feel a little anxious about my work I take a pen."

With female typists in the closest proximity to those giving them dictation, the workplace became newly sexually charged. It took some getting used to—for both employees and their families. In Atlantic City in 1892, two high-profile court cases revolved around men who had married their "typewriters"—there was no distinction, apparently, between the women and the machines they worked upon. Relatives of two different men who had married their typewriters attempted to annul the marriages, stating that the men had not been of sound mind when the marriage had been entered into.

The Great Postcard Craze

Oddly, the change in writing practices that had a greater impact on what people wrote to one another in private was a small, square piece of card—the *carte de visite* or, as it soon began to be called, the postcard. Rumor had it that the thing had been invented on the Left Bank in Paris, where a man spilled his coffee on a square piece of writing stock. The stain made an interesting shape, so he affixed a stamp and an address to it and mailed the card to a friend. In America, another story revolved around "an economical young woman in San Diego who had to pay postage to write her sweetheart, but who would not buy writing paper. She wrote her epistles in minute penmanship on the reverse side of a stamp and mailed only the stamp itself."

In any event, the first postcard was sent in England in 1871, and by 1873 more than 72 million of them per year were dropped into the British post. That same year, 26 million were sent in Germany, which later became the nation that printed most of the world's postcards. The postcard craze had arrived. It's easy to see why people took to it. A postcard was a cheap, relatively quick way to say, "Yes, I received your letter"; to send and receive, accept and decline invitations. Doodles, jokes, and romantic asides traveled this way at a fraction of the cost of a telegram.

In the era before cameras were portable and cheap to own, postcards allowed tourists to bring back some sort of visual rendering of where they had been. As Susan Sontag has noted, this newly created act of recording what had been seen led to an intellectual idea that all the world's visible things—be they a painting or a pizza—could be captured on film and, later, that the purpose of the travel was the obtaining of that image. Before this development, the way people recorded was that they remembered, or they sketched, if they had the inclination, or kept a private diary. What is unique about all of these activities is that each has a singular aura. Postcards, however, were mass-produced.

That didn't bother most travelers. By the early 1900s, the postcard had become a full-fledged obsession in America as well. In 1906, it was estimated that one in eight Americans bought a postcard every day. The country spent more than $1 million on the little pieces of stationery each week, and in the course of just a few years they became available at more than eighty thousand merchants nationwide. Many of these stores began selling albums for collectors, which ranged from less than a dollar up to $15. Postcard clubs, which allowed people to trade postcards from faraway places, sprang up.

Postcards also became a way to teach geography. One New York state schoolteacher in a small rural town with no public library started a pen pal course between her pupils and foreign students in Africa, Australia, Ceylon, Cuba, Iceland, New Zealand, most of the countries of South America, and all of Europe. "The postcards brought children into touch with the whole world in a way no other means at their command would have done," wrote a newspaper reporter covering the story of the children's correspondence. Some places they received return cards from didn't even appear in their textbooks.

But the little invention was not without its abuses. A *New York Times* article in 1871 reported on behavior that resembles a print version of flaming (the practice on the Internet of harassing and criticizing someone publicly):

> The handy little post-card has already been made the instrument of insult, ridicule or revenge. The anonymous letter has always held to be one of the most cowardly weapons of assault ever used among civilized beings; but with such letters the sting was limited in its application. The receiver might suffer, but he had the option of suffering alone. . . . The post-card, however, enables concealed scoundrels to deprive their victims even of this discretion. Gross insolence or contemptuous epithets can now by this means be leveled. . . . We regret to observe that many such instances have lately occurred in London, so many as to constitute a grave objection to the post-card system altogether. . . . The temptation to call people liars and scoundrels in so safe a way, to accuse them of robbing hen-roosts or murdering their grandmothers, seems quite irresistible to many ingenuous souls, and

impunity has apparently brought about something of an epidemic.

Swindlers, Snake Oil Men, and Peepers

Postcards and typewritten letters highlighted a growing problem with written communication: Can you trust the person on the other end of the line? Handwritten letters had their own signature of authenticity, but typewritten letters instantly created a different impression, one of professionalism and business-mindedness—a fact exploited by some aspiring men, as revealed in this newspaper story:

> The other day a lawyer had just finished a letter on his typewriter with the word "dictated" at the bottom of it. "Why did you add that to it when you wrote it yourself?" asked a friend. A look of pity filled the lawyer's face at the stupidity of his visitor. "My guileless, far-away correspondents," he said, "will believe that I am overrun with business and utterly unable to answer my own letters. If they regard it as a luxury for me to have a private secretary, why should I undeceive them?"

The growing lack of face-to-face communication also presented an ideal situation for all kinds of criminal or miscreant schemes. The earliest adopters of new modes of communication are often those engaged in illegal or unethical activity, who benefit greatly by being ahead of the curve. The newspapers of the 1880s were full of stories about swindles and heists perpetuated by the mail.

In 1884, an advertisement was placed in several Brooklyn papers: "ONE THOUSAND DOLLARS will be paid for information identifying the author of certain anonymous letters mailed to residents of the Nineteenth Ward during the last two weeks." The advertising copywriter should have been more specific. In fact, a batch of Valentines, some "written in a woman's angular hand," some composed by typewriter, had been sent to women around Brooklyn containing messages so lewd that *The New York Times* couldn't print them. Chaos had ensued. "In two families the marriage engagements of daughters have been broken off through the instrumentality of these letters."

In most cases, financial gain was the goal. A New York man procured a list of lumber dealers and opened correspondence with them. Letters to his bankers in Philadelphia to verify his creditworthiness would "bring the reply that Mr. Rowe was a fair, honorable business man." No sooner had the lumber been delivered, though, than he vanished. In the course of an investigation, an expert on the typewriter was called in to testify that the notes and the bank correspondence were the work of one man.

One of the most successful schemes of the day will ring bells for anyone who has received e-mails from a Nigerian attorney promising hidden millions. In 1887, two British men set up a business called the British-American Claim Agency in New York City, writing to people around the country and encouraging them to look up claims to estates of long-lost relatives in England. The London *Times* was made to appear an endorser of the scheme, and to light a fire under their "heirs," the amount the paymaster in Chancery had ready to deliver to them was, according to *The Times,* £77,693,769.

All the targets had to do to start claiming their rightful fortunes was to write to the office and pay a $2 fee. Five dollars secured an advertisement in London and $13.25 a guaranteed search. Police in the United States were tipped off to the heist by a judge, who shared the same downtown office building as the men behind the scheme; the two men, he said, had been receiving an unusual volume of mail. At the miscreants' office, law enforcement discovered that the two men had employed fourteen young women to type up the advertising circulars on which the British-American Claim Agency sent out its entreaties. In reality, no searches had ever been performed, complaints were ignored, and the two hauled in as much as $500 per day.

Finally, as more and more people communicated by letter, a larger question loomed: Could governments be trusted—as they could not in the early days of mail—to preserve the privacy of the post? Writing in 1960, Summerfield was adamant about the sacredness of first-class mail: "Its privacy is zealously guarded from the moment it is mailed until it is delivered. Not even the President may order it to be censored, or delay its delivery, except in time of war." This was a convenient caveat for a country regularly at war, as the United States was from late in 1939 through the 1960s and is again today.

In April 1976, a U.S. Senate panel discovered that the FBI, CIA, and several other government agencies had illegally monitored millions of telegrams and opened more than a quarter million letters. Presidents from Franklin D. Roosevelt to John F. Kennedy were implicated. And it was not just a simple scan of letters. As Seymour Hersh reported, James Angleton, the longtime head of the CIA's counterintelligence division, was personally involved in a "series of (illegal) domestic mail intercepts that enabled the agency to learn how the American Federation of Labor was planning to use the millions of dollars in clandestine funds funneled to it by the C.I.A." According to one account, "Angleton would personally deliver copies of the letters to Allen Dulles—and thereby 'made real hay with Allen' since 'it impressed Allen enormously to know in general' what the AFL was planning to do."

The Plague of Words

Swindles and con men—and the occasional government peeper—were not the only problems with the post as the world tipped into the twentieth century. There was also the sheer amount of it. Not everyone was pleased by the large numbers of letters and postcards flying across the plains. At the turn of the twentieth century, worries over the decadence of the age of letter writing popped up in opinion articles and among men and women who saw themselves as protectors of civil society. "There is no standard nowadays of elegant letter writing," said one woman, "as there used to be in our time. It is a sort of go as you please development, and the result is atrocious. Epistolary accomplishment is considered altogether too puerile a study for the strenuous work of higher education, while rapid note taking at lectures, etc., finishes the ruination of handwriting and style, the result being as you have just observed—that our daughters write like housemaids and express themselves like schoolboys."

The postcard was often blamed for this drop-off. "It has frequently been remarked during recent years that the art and practice of letter writing has passed away, and the picture postal has helped on this tendency," wrote one correspondent in *The New York Times*. "People write less than they ever did, and yet they keep their friends at home posted as to their itinerary during a long trip better than ever before. The picture postal tells a story. That is why it is so popular." As a result, people heard from one another a lot more, especially when they traveled on vacation. "People will send tourist postcards when they would not write letters."

Letter-writing mavens in newspapers turned their attention away from simple epistolary etiquette and toward the more pressing problem of the letters one absolutely had to send. "Certain letters, however, must be written," opined Mrs. Van Rensselaer Cruger, writing under the pen name Julien Gordon; "there is no escape from their claims. These inevitables may be classed as follows: The Family letter; the friendly letter; the business letter; the letter of condolence or felicitation; the love-letter; the miscellaneous note."

Even President Theodore Roosevelt got into the act of scolding people for their prolixity. "A resolute effort should be made to secure brevity in correspondence and the elimination of useless letter writing," he argued in November 1905. "There is a type of bureaucrat who believes that his entire work and the entire work of the government should be the collection of papers in reference to a case, commenting with eager minuteness on each, and corresponding with other officials in reference thereto. These people really care nothing for the case, but only for the documents in the case. In all branches of the government there is a tendency greatly to increase unnecessary and largely perfunctory letter writing."

An unsigned commentary appeared soon after Roosevelt's scolding comments, applauding his appeal to people's sense of restraint. It's worth quoting at length for the way it captures the sense of fatigue, fatality, even, Americans felt when facing the future of words:

> We hope the President will be a restraining influence
> on the flood of words both in correspondence and in

> books, but we fear the times are against him. They offer fatal facilities for verbal exuberance. Books today are published in vast numbers, less because authors have anything to say than because printing is easy and cheap and the presses have to be kept at work. So, too, the typewriters click out folio after folio in public offices, not because there is any real reason for that amount of writing, but because the machinery for producing it is at hand. . . .
>
> The stenographer, the typewriter and the printing press are invaluable agents of civilization, but they have their drawbacks. They have inundated us with a plague of words, and we wish that curtailment in the government service could be but the beginning of reform.

In some cases, though, it wasn't just bureaucrats adding to the blizzard of words. By the end of the nineteenth century, commentators began to complain about being badgered at home by news about political candidates. "Even the Republicans favor me with a tableful of campaign documents, possibly to keep me strong in faith," wrote William Drysdale in a piece entitled "Does Anybody Read Them?" in *The New York Times* on November 6, 1887. "It occurs to me that maybe these soul-stirring papers are sent to all the good Republicans, in hope that they may feel flattered by the little attention; but I am never flattered by the receipt of anything short of a Patent Office Report, or a bound volume from the Department of Agriculture. As I sweep all these documents into the waste basket with one grand swoop, they inspire me with only one thought. It is—'How the printers must smile, when they see an election coming on.'"

The Eighty-Letter Day, the Power, and the Glory

The basic postal rate—at least in the United States—didn't go up for another fifty years once it was first lowered, and even then it rose by just a cent. The era of postal overload was here to stay. Aside from businesses and governments, people who had come into a peculiar kind of modern existence—being famous—weathered this overload in exaggerated fashion. The prolific journalist and essayist H. L. Mencken felt duty-bound to respond within the same day out of "decent politeness." He also followed the do-unto-others rule: "If I write to a man on any proper business and he fails to answer me at once, I set him down as a boor and an ass." Therefore, every day, whether the mail brought ten or eighty letters, he read and responded to all of them. "My mail is so large," he said, "that if I let it accumulate for even a few days, it would swamp me."

H. L. Mencken, possibly answering a letter

Others had someone close by to do the answering. Thomas Edison received thousands of unsolicited letters per year and employed a fleet of male secretaries to craft his responses and sift the nuts from the fans. After winning the Nobel Prize in 1930, the American novelist Sinclair Lewis began to receive hundreds, including several appeals for help or employment. "I'll do everything for you," one woman wrote, "and when I say everything I mean everything." "My dear Miss," replied Lewis's wife, Dorothy Thompson, to her husband's admirer. "My husband already has a stenographer who handles his work for him. And, as for 'everything' I take care of that myself—and when I say everything I mean everything."

But an even bigger problem had yet to be confronted: junk mail.

The Business of Moving Mail

When Henry Raymond unleashed *The New-York Daily Times* upon the city, he appealed to New Yorkers by letter:

> The carrier of "The New-York Daily Times" proposes to leave [the newspaper] at this house every morning for a week, for the perusal of the family, and to enable them to receive it regularly. The Times is a very cheap paper, costing the subscriber only SIXPENCE a week, and contains an immense amount of reading matter for that price. . . . It will contain regularly all the news of the day, full telegraphic reports from all quarters of the country, full city news, correspondence, editorials. At the end of the week the carrier will call for his pay;

and a continuance of subscription is very respectfully solicited.

As people used the world's emerging postal services more often, business began to use it, too, in order to target customers. The rise of direct-mail solicitations was fast and had been part of life in the colonies from the beginning. The U.S. Postal Service has always operated second-class mail, through which magazines, newspapers, and advertisements travel at a loss, the reason being "that a postal system should help disseminate information as a public service and do so," Arthur Summerfield observed, "partly at least, at public expense." *The Federalist Papers,* essays on articles of the Constitution written by James Madison, Alexander Hamilton, and others, traveled this way, in the pages of the New York newspapers *The Independent Journal, The New York Packet,* and *The Daily Advertiser,* which in those days arrived by post.

But how much can be classified as information in the public service? Apparently a lot, and it took an energetic British man to truly exploit that wide definition. The advertising circular, which traveled in second-class mail, was invented by G. S. Smith in London in a borrowed office in 1868. Smith was just fifteen, and he addressed all the pleas for purchase by hand. Smith used halfpenny wrappers, owing to postal regulations. Within a short while, he had several men working for him and then an army. Before long his company could issue prospectuses for publicly traded companies in London (1.25 million copies for the Manchester Ship Canal Company) and in America (2.5 million copies for an American finance house). In the early 1900s, he was the world's biggest purchaser of envelopes, one of his orders clocking in at over 100 million of just one kind of envelope. By the time he died, he employed more than 300 men and 130 "girls." All of the letters sent out were addressed by hand.

Businesses began to solicit customers by catalog in the middle of the nineteenth century, the first produced by Aaron Montgomery Ward in 1872. Around the same time, companies began selling typewriters by mail. The huge increase in advertising mail led to the U.S. Post Office running a significant budget deficit at the turn of the twentieth century. Marketers got hold of mailing addresses and batched them into groups that could be sold. In December 1913, the president of the Kentucky Distillers Company offered to sell a mailing list of fifty thousand customers—"each individual on the list is a regular user of liquor"—to a Kansas City, Missouri, sanatorium for alcoholics. The Anti-Saloon League then duplicated the letter in a leaflet to show the lengths to which the greedy liquor industry would go to take advantage of its customers.

In the 1930s, looking to bolster its flagging earnings in the Great Depression, the U.S. Post Office began encouraging advertising mail, effectively putting the government into competition with the nation's newspapers (which couldn't function without advertisements). "If successful in any large way," complained Eugene Meyer, the publisher of *The Washington Post,* in November 1934, "[the U.S. Post Office's campaign] would naturally reduce the legitimate receipts of the daily newspapers of America and thereby weaken their position." Third-class bulk mail rates were introduced by the United States in 1928, and selling took off.

The phrase "junk mail" first appeared in 1954, and people began to fight back. A Connecticut man, irritated at the state's avowed practice of selling lists of its registered drivers and all the junk mail he received as a result of it, refused to tell the state's Department of Motor Vehicles his new address. The list upon which the man's name appeared had been sold to R. L. Polk & Co., a Detroit marketing firm, for $15,000. The company had been buying lists of registered drivers from all fifty states for thirty years. The man lost his case, and junk mail has proceeded apace. In 2003, 43 percent of U.S. mail was direct mail, up from 29 percent in 1980.

Electronic Brain to Sort Mail

As the number of pieces of mail entered the billions, post offices around the world began to creak under the pressure. The Canadian Post Office Department became the first to invest in electronic sorting machines. A scientist, Dr. Maurice Levy, was the first man to perfect such a machine, and it went into service in Ottawa, Toronto, and Montreal in 1957. The United States, which by 1960 had just one-fiftieth of the world's population but two-thirds of its mail, was not far behind. Machines were installed in major cities that made huge gains in sorting time: an electronic machine could sort 21,600 letters an hour, compared with the 1,500 managed by a good postal worker. Delivery times were speeded up by just 50 percent, though, since the biggest difficulty for the mail wasn't in the sorting but in getting it into and out of the eleven major metropolitan offices through which two-thirds of all American mail traveled.

Postmaster General Summerfield, however, was not to be deterred. "I will not be satisfied," he said in 1957, "until we can give patrons delivery of letters between any two American cities on the day after mailing." The loftiness of this goal explains why, from the early 1960s on into the 1980s, the post office began to invest in a peculiar solution known then as electronic mail—today we simply call it a fax. In 1961, in Washington, Chicago, and Battle Creek, Michigan, a service was tried out through which correspondents sent an electronic message; on the other end it was printed out and delivered as a regular piece of mail. The service was dismantled in the early 1960s, then tried out again at the end of the decade between Washington and New York City. The experiment was again short-lived due to lack of patronage.

The post office kept trying, though, as telex machines and the fax began to eat into its market share. A new era of communication was coming, and given that it was already being subsidized at a rate of up to $2 billion a year by the U.S. government, the post office couldn't afford to be behind the curve: early estimates suggested that 17 billion pieces of mail could be electronically redirected by the mid-1980s. In the early 1980s, the post office spent close to a million dollars trying out the "electronic mail" solution between two American cities and several European countries. It was refined and labeled E-COM, "a special service for businesses with a sufficient volume of mail and enough computer capacity to take advantage of it," as described in *The New York Times*.

To pull it off, the post office turned to an old player in the communications game: Western Union. Organizations that sent large volumes of mail would transmit computer-generated messages over Western Union lines to twenty-five important post offices. Upon arrival, the messages would be printed out and put into envelopes, with the rate at thirty cents for customers who sent fifty thousand letters in four weeks and fifty-five cents for customers who sent only five thousand messages in the same period. It was a bold step into the future—but it was ultimately not to be, as the post office kept running into one obstacle: the U.S. government.

Time and again the post office's electronic mail schemes ran into opposition from the Federal Communications Commission, which unanimously blocked its first official attempt to enter the electronic mail age, citing a lack of data from Western Union, which had applied for a license to carry mail over wires on the post office's behalf. The FCC also believed that since the mail had been sent electronically, it fell under its jurisdiction, not the post office's. Later the Justice Department, the Commerce Department, and senators beholden to large industry fought against the scheme. The program was short-lived. Begun in 1982, it sent 16 million messages per year; it was folded in 1985. As a result of its failing, today we can send e-mail without a stamp.

The Ultimate Destruction of Space

As the post office discovered, systems, especially those that deal with the shipment and transportation of tangible objects, have a terminal velocity. Conveyor belts can spin only so fast before the objects they're transporting fly off. Trucks and buses have to obey speed limits; aside from the now-defunct Concorde, commercial airliners' speeds have remained constant for decades. Mail that is electronic on one end and physical on the other—as E-COM was, its regulatory issues aside—solves only half the problem. The second a letter was printed out, it crashed into the bedrock of reality and slowed to a crawl.

In order for terminal communicative velocity to be reached, however, time and space didn't just have to be destroyed, they needed to be reinvented. Mail, by closing the gap between California and Connecticut, let alone Calcutta and the Cotswolds, began to bring about this change. The written word became an intimate tool that everyone could use, and as a result the sphere of intimacy expanded. But the lightning bolt that truly changed our sense of time was the telegram, which didn't just speed up words but instituted a kind of new reality—one that found an echo in the burgeoning newspaper industry, on battlefields, and in that mundane symbol of modern man: timepieces.

2

THE INVENTION OF NOW

Had there been stretched across the Continent yesterday a line of clocks extending from the extreme eastern point of Maine to the extreme western position on the Pacific coast, and had each clock sounded an alarm at the hour noon, local time, there would have been a continuous ringing from the east to the west lasting for 3¼ hours. At noon today, there will undoubtedly be confusion.

—*The New York Times, 1883*

On November 18, 1883, one man stopped time in New York City for nearly four minutes. The fellow thumbing the watch springs to a halt was one James Hamblet, the general superintendent of the Time Telegraph Company and manager of the time service of Western Union. In this capacity, Hamblet was effectively Gotham's archduke of time, a role he had earned through hard work and creativity. Hamblet had invented an electric clock that could chime in a remote location, a device of great use for railway stations, which were required to display the time. Hamblet also managed Western Union's own finely calibrated clock in room 48 of its 195 Broadway office. On that day, the regulator, as it was called, kicked off the mammoth task of synchronizing railroad timetables—no small feat, since as late as 1882 American railroads had a blizzard of time standards and therefore possessed more than seventy different answers to one very simple question: What time is it?

195 Broadway

Hamblet's was not as dangerous a juggling act as one might think. Even though early American rail lines were constructed to travel on a single track, a small glitch in scheduling would not send a huffing Yellowstone Park Line crashing into a Northern Pacific waiting at the station. Telegraphic control of train movements, which began around 1855, prevented such accidents. Before that, complicated timetables invented by the French engineer Charles Ybry kept the rails safe.

Still, passengers and station agents constantly wrestled with a persistent irritation: railroad time was often slightly different from local time—even more so outside major cities. As a result, "any traveler . . . upon leaving home, loses all confidence in his watch and is in fact without any reliable time," wrote Charles F. Dowd in 1869. If a passenger planned to travel from San Francisco to Washington, D.C., he would have an even more niggling problem: to keep up with local time, he would have to change his watch more than two hundred times along the way.

In the middle of the nineteenth century, the converging needs of geophysics—for uniformity of observations—and railroads led to a syncopated, haphazard, but effective push to fix this situation. In January 1882, Professor Cleveland Abbey, at a meeting of the New-York Electrical Society, proposed three standard times: Philadelphia time for the Atlantic coast, Saint Louis time for the Mississippi Valley, and San Francisco time for the Pacific coast. In October 1882, the heads of all the major railroads met in Chicago, where they agreed to work together to create standardized time.

A year later, at precisely 9 a.m. in New York, Hamblet stopped the regulator for 3 minutes and 58.33 seconds—so that he could standardize time to a reading taken from a nearby observatory—and then restarted the machine, creating a new 9 a.m. sharp. Three observatories—in Washington, D.C., Cambridge, Massachusetts, and Allegheny, Pennsylvania—then tested its accuracy by telegraph. Finally, at noon, a ball dropped from the top of the Western Union building, which triggered a telegram to be sent to the city's more than two thousand jewelers, who, in addition to peddling diamond broaches and pearl chokers, sold time itself.

It is here—at the jewelers'—that we get a fascinating window into the metaphysical vertigo that overcomes us when the space-time continuum is disrupted, sped up, or stopped altogether. On a small scale, November 18, 1883, sounds like a Y2K of the nineteenth century. Many New Yorkers who wandered into jewelry stores that day seemed to think that the hiccup in their clocks "would create a sensation, a stoppage of business, and some sort of disaster, the nature of which could not be exactly ascertained." Storefronts did not flog duct tape or bottled water, but a similar letdown descended upon the befuddled when the fateful hour passed without catastrophe. "They were incredulous when informed that the change would probably be one which they would know nothing about at the time," wrote a *New York Times* reporter in a story entitled "Time's Backward Flight," "and would not necessarily postpone the celebration of Evacuation Day for a week." Shipmasters, arguably, faced a more practical problem: they would have to figure out how to coordinate their position in this new linked scheme when "sailing about out of the reach of time-balls."

All Together Now

The way Hamblet and company went about melding railroad times brings back a lost world, one that seems quaint in our age of atomic clocks and handheld satellite navigators—we who always know exactly where and when we are, even if the road runs out. November 18, 1883, also highlights a truth that undergirds the grid of technology upon which modern life depends: all major new technologies affect our sense of space and time, and any technology that alters these elements also alters communication. The faster we relay information and the more we share what goes on in our heads with others, the busier our society becomes—space that was once thought conquered, such as the vast stretches of the American West or the wide blue deeps of the oceans, reassert themselves into virtual spaces, which become crowded as people regroup, and what is shared within them approaches unmediated human thought.

The state of frenzy in which we live now was a long way off in the nineteenth century. People lived under the same darkening sky, but they did not live simultaneously. This is an important distinction to contemplate today, when so much of what we do—and especially what we communicate to one another—depends upon simultaneity. We wouldn't have a media age without it. Everything from watching a television show broadcast out of New York while sitting on a couch in Chicago to sending an e-mail from one computer to the next to coordinating travel plans on the Eurostar could not happen without an agreed-upon sense of what "now" means. We could not travel by airplane or perform scientific experiments or trade stocks online or even clock into and out of work without it. Scientific experiments were, of course, completed in the nineteenth century and earlier, but the lack of standardized time was a constant stumbling block. And the now that we live in today—the now that many of us experience most intimately through the daily onslaught of e-mail, which has quickly developed a culture and expectation of instant response—has important roots in, but is vastly different from, the now people the world over were trying to wrap their heads around in 1883.

At the fin de siècle, people around the world—let alone in the next town over—did not occupy an agreed-upon sense of time and place. They lived on a multiplicity of slightly different schedules. The International Prime Meridian Conference held in Washington, D.C., in October 1884 set up a single prime meridian passing through the Royal Observatory at Greenwich

(Greenwich Mean Time) and adopted a twenty-four-hour day. But to the great disappointment of Sandford Fleming, a Canadian inventor, builder, and railway engineer, the meeting did not succeed at establishing standardized time for all nations. England was ahead of the game: it had used a standardized time system, based on Greenwich Mean Time (GMT), for the railroads since 1847. Most English clocks were synchronized to GMT during 1855. France and Spain did not follow until the early twentieth century.

For most of the rest of the world, until the twentieth century, time was kept by the solar noon. It seems like an organic solution, but it was not very helpful for unifying nations, let alone cities. The sun keeps moving and the earth keeps revolving, twelve and a half miles per minute, which means that every twelve miles experiences a different noon. It wasn't just that Cleveland was different from Calcutta, and Detroit was different from Denver; *neighborhoods within the same city* were on different timetables. "Adjacent villages clung jealously to their particular time with all the ferocity of a threatened identity," wrote Clark Blaise in *Time Lord,* "accusing each other of keeping false time." This myriad of time zones reinforced the nature and importance of distance; the context of life was local. People knew the earth was round, the oceans regulated by the moon. But most of the rest of the world was truly *elsewhere.*

This was not a confusing experience until two of the nineteenth century's most powerful technological forces, the railroad and the telegram, combined. For the first time in the history of humankind, people on opposite sides of the globe could communicate almost instantaneously, and it was the railroad—which allowed them to get there eventually—that had made this possible, as telegraph lines traveled along the grid of physical tracks. Western Union, which was incorporated in April 1856 and by 1870 held a monopoly on U.S. telegraphic communication, franchised its offices out to station houses employing several thousand operators. When Samuel F. B. Morse sent his famous first telegram from the Law Library of the U.S. Capitol all the way to Baltimore, it hummed along the Baltimore and Ohio Railroad's rail line. The very second message sent was HAVE YOU ANY NEWS?

Knitting Nations Together with Words

We human beings are known for transforming—and adapting to—our environments, but adjusting to a frame of reference created by electronic communication amounted to a vast change. After all, the speed at which words travel has implications not just for commerce but for how we mark our place in the world. It is hard to underestimate the way these new communications links helped to create a sense of nationality, especially in the United States. After all, it was here that the physical expansion of a nation—the gradual diminishment of "wilderness" by virtue of imperialist expansion into Indian territories, the connection of one state and city with the next by way of railroads—was inextricably linked with the country's sense of time. It suggests that, whatever philosophers argue goes on in our minds, time and space do not exist independently in a body politic; they must be woven together to create a kind of national reality. Scientists the world over had been pressing for synchronized time. But it was the railroads and the burgeoning telegraph empire of Western Union, those two early monopolies at the heart of American power, that drew the country into a simultaneous now.

The hopes for what this new now would bring were enormous. It aroused feelings of patriotism and, more important, a desire for unity that would prove a dream in the short term, with the Civil War around the corner. As one observer noted in 1846, "The power of the States will be broken up in some degree by this intensity and rapidity of communication, and the Union will be solidified. . . . We shall become more and more one people, thinking alike, acting more alike, and having one impulse."

The U.S. government was slow to embrace this sentiment, however. The telegram faced enormous skepticism in a world where the discovery of electricity still felt like a conjuring trick. Samuel F. B. Morse began plying the U.S. government for funds in 1838 and became so frustrated he traveled to Europe to drum up support, unsuccessfully. In England, the inventors William Cooke and Charles Wheatstone had been peddling their own electric telegraph to the English government with equal lack of success. They eventually convinced the Great Western Railroad to run a thirteen-mile telegraphic link between Paddington and West Drayton, and Blackwall Railway in London's docklands. Their successful installation and use of the telegraph to catch robbers—descriptions of whom were sent down the line, aiding in their apprehension—eventually won people over, and Cooke enjoyed an enormous public relations coup when one of his early lines was used to announce the birth of Queen Victoria's second son, Alfred Ernest Albert, on August 6, 1844.

In the United States, a bill was finally put forward to fund Morse's proposed line in 1842, granting him $30,000 of federal money to develop his telegraph, but not before Representative Cave Johnson of Tennessee goaded the House, which had not funded many experiments in science at that point, into considering giving half of the earmarked funds to Theophilus Fisk, a proponent of mesmerism, also known as hypnosis. Fisk's earmark did not pass, but spiritualists would later petition the U.S. Congress again on the belief that in the mysterious workings of the telegraph there might be hints of a much greater "spiritual telegraph" between Heaven and Earth.

Morse responded as a scientist; he would show that his ideas could be verified factually. As he began building his first line, he visited the Capitol to demonstrate how his telegraph worked, hauling a five-mile-long cable into the building to show messages being passed along it (through a system of electrical pulses, which were then decoded on the opposite end). No one understood what this experiment proved. Morse finally convinced his skeptics two years later when, in May 1844, he sent word that James K. Polk and Henry Clay were nominated for the presidency by the Democratic National Convention from Washington, D.C., to a temporary platform fifteen miles outside Baltimore, beating a courier sent along the railroad by nearly an hour.

Morse has since been credited with inventing the telegraph, but he merely perfected an idea that had been in development for nearly a century. The word "telegraph," which literally means "that which writes from a distance," was coined by the French, who pioneered the development of the optical semaphore system that predated the electric telegraph. Morse's incarnation was quickly picked up and developed by state governments throughout Europe. The Belgians built a line connecting Brussels and Antwerp in 1846, while the French opened their system to the public on November 29, 1850. In the Netherlands, a royal ordinance of December 8, 1847, made private telegraphic undertakings so circuitous as to keep telegraphy practically a government operation. The Holland Railway Line, however, opened its own services to the public in 1845, and thereafter it fell to cities and municipalities to build their own lines connecting to the trunk system built by the central government. Prussia and Austria followed in 1849; Bavaria and Saxony in 1850; Sweden in 1853; Denmark in 1854; Norway, Spain, and Portugal in 1855; Russia in 1857; and Greece in 1859. All of the important lines in these countries were built by the government, and by 1904, the United States was the only country where the telegram wasn't publicly owned and operated.

In the United States, it was primarily the private companies, the railroads, and the military that laid the telegraph lines. The Civil War of 1860 was only the second major world conflict (after the Crimean War) in which generals talked to their troops and field commanders via a communication network that differed significantly from that used by generals two thousand or three thousand years previously. Union and Confederate troops laid wires on their marches north and south, establishing some fifteen thousand miles of the network for military purposes. Three hundred operators died while sending messages.

Civil War telegraph wagon, 1864. *Photo by David Knox.*

President Lincoln, much as Obama would later be with e-mail, was attached to the machine at the hip. "For the last two or three weeks of his life Lincoln virtually lived at the telegraph office," sociologist Robert Rupp has written. "The wires were kept busy with dispatches to and from the President." He would often peer over the shoulders of the cipher operators when an important message came in and was being decoded. He used the telegram to urge his generals on from afar. I HAVE SEEN YOUR DISPATCH EXPRESSING YOUR UNWILLINGNESS TO BREAK YOUR HOLD WHERE YOU ARE, he wrote to General Grant in City Point, Virginia. NEITHER AM I WILLING. HOLD ON WITH A BULL-DOG GRIP, AND CHEW AND CHOKE, AS MUCH AS POSSIBLE.

The opposing armies occasionally even used the cable to talk to each other. I SEE YOUR CONDITION THROUGH MY TELESCOPE, wired Confederate General Pierre G. T. Beauregard, after seeing a truce flag waved by a Union leader, Robert Anderson. WE HAVE INTERCEPTED YOUR SUPPLIES. GIVE IN LIKE A GOOD FELLOW, AND BRING YOUR GARRISON TO DINNER, AND BEDS AFTERWARDS. NOBODY INJURED, I HOPE? Thanks to the telegraph, news of their victories and losses appeared in newspapers the next day.

News from Afar for Everyone

Once information of national import could be conveyed by telegram, it created a new media landscape. Until this point, newspapers had had to rely on the speed of couriers, rowboats, and the Pony Express to beat one another to breaking a story. In one fell swoop, the telegraph fundamentally changed how newspapers competed for customers, since any organization with access to a telegraph could get access to news. In the 1840s, the *Times* of London—which regarded itself as a global newspaper, in keeping with England's far-reaching empire—carried news from New York that had taken five weeks to arrive, from South Africa that dated back seven weeks, and from India that was nearly two months old. By 1870, after the laying of the transatlantic cable, which allowed telegraphic communication between America and Europe, that Doppler wave of news was eliminated, and not just for Londoners. A doctor in South London could pick up a newspaper and read of events that had happened in South Dakota *hours* before. As telegraphic cables were laid from one continent to the next, a baker in San Francisco could learn about floods in China, a fire in Bombay, and the birth of a royal offspring in Austria.

Newspapers' reaction to this development was simply to print bigger editions in order to accommodate all the news. In Britain, this was made possible by the 1855 repeal of the Stamp Act, which had made newspapers prohibitively expensive. In fact, Lloyds Bank in London, which had begun as a coffee shop and briefly printed its own newsletter, even had an appointed time when, upon the ringing of a bell, a man would stand up and read the morning's newspaper aloud so that its patrons could share the expense.

Between 1855 and 1870, a large number of biweekly newspapers became dailies, including *The Manchester Guardian* (1855) and *The Glasgow Herald* (1859), rearranging the media environment. London papers began to circulate around the country, and since they had to go to press earlier to get onto trains, news in the Glasgow and Manchester papers was actually three hours fresher than that in the capital. The beginning of newspapers' great age was upon the UK. In fiscal year 1869–70, Great Britain had 1,450 newspapers with an aggregate circulation of 350 million.

Telegrams were essential to this growth. In England in 1872, the telegram ferried 40 million words of newspaper dispatches, constituting one-fifth of the volume of all telegrams sent, the equivalent of forty entire sets of the *Encyclopaedia Britannica*. Their contents were typed out and passed along by hand. A similar effect was observable across the English Channel. The number of French newspapers exploded after the 1830 revolution, with an "avalanche threatening to overwhelm Paris," and by 1845 there were more than 65 million copies in aggregate circulation. In 1847, fed by telegraphic links to the world, Paris alone supported twenty-six daily newspapers, with a strong demand for literary matters by literary readers.

The American newspapers had a head start on this golden era of public information, as newspapers had been intimately involved in both the War of Independence and the discussion surrounding the First Constitutional Congress. Indeed, several founding fathers—including Benjamin Franklin, who edited one of the nation's first newspapers, and Noah Webster, who founded New York's first daily newspaper, *The American Minerva,* at the behest of Alexander Hamilton—were ex-journalists of a sort. To be a newspaperman, in a sense, was to be in touch with one's people, an idea at the heart of American democracy.

In 1800, there were 376 newspapers in operation in the United States; by 1835, that number had quadrupled to 1,200; and by 1870, there were 5,871 newspapers with a total circulation of 20 million copies, and newspaper circulation was increasing by 15 percent a year. Many of them were cheap, and they were packed with news. The early telegraphic companies briefly tried to sell news to the papers. As it turned out, the opposite happened. In *The History of American Journalism,* James D. Startt and William David Sloan chronicle how newspaper editors banded together and "created a system of newspaper-generated and newspaper-owned information sent to other newspapers over leased wires." In 1848, as war with Mexico loomed, the

Baltimore *Sun* joined the *New York Herald* and several other newspapers to cover the Mexican War, the first American conflict that received nearly same-day coverage in the press. Out of this collaboration was created the New York Associated Press, a cooperative news agency that sold news to papers throughout the country. This arrangement soon spread to other parts of the country, until in 1900 the Illinois Associated Press reincorporated in New York and henceforth became known as the AP.

Meanwhile, something slightly different was happening in Europe. The German-born Paul Julius von Reuter, who until the 1830s had run a translation house and a service that sent messenger pigeons with closing stock prices to a central office, began to offer news—mostly financial—by telegram. For the first year, he sent the pigeons as well, as backup, using them to bridge the gaps between nations that did not have telegraphic treaties. Gradually, he expanded out of financial news, proving how extraordinary his service was in 1859, when he "obtained a copy of a crucial French speech concerning relations with Austria and was able to provide it to the *Times* in London within two hours of its being delivered in Paris."

All across Europe, a new class of everyday readers was being created. Newspapers helped lead this group of readers to the newly popular form of the novel, running book reviews, especially the newly expanded "provincial papers," such as *The Manchester Guardian* and *The Scotsman,* which had a strong driving ethic of social reform and were therefore especially interested in the circulation of ideas. Lists of new titles that had appeared in local booksellers appeared in these sections, while serials of emerging works by Charles Dickens and other writers appeared in the weekly versions of daily papers and were syndicated by authors to fifteen or twenty papers, sustaining, in piecemeal fashion, the audience for the flourishing genre of the novel.

Newspapers were soon filled with other news that the telegraph made possible: weather forecasts, synchronous scientific observations, speeches by politicians, up-to-date stock prices from markets around the world, descriptions of foreign battles. For a brief moment, thinking it would encourage patriotism and support for the Crimean War, London newspapers made the mistake of printing information about troop and fleet deployment, much to the delight of foreign spies in the city.

The telegraph did not just institute a one-way communication from newspapers to the people; people talked back through it, as well. When the International Workers of the World threatened a walkout of 250,000 men—miners, harvest hands, and lumbermen—if deported workers were not returned to their home in Arizona, they sent their final demands to President Woodrow Wilson by telegram. A few years later, the Emergency Peace Federation staged a telegraphic protest to the U.S. buildup to World War I, flooding newspapers and representatives with a million messages of peace. It made such a stir that the Naval Training Association urged its two thousand members to send their own cables in support of military action.

The First Era of Information Overload

Managing the onslaught of the world's muchness, brought to people by virtue of the telegraph, quickly became a quandary for individuals, especially businessmen. Merchants who depended on commodity prices had to check fluctuations with increasing regularity, and their monitoring and manipulating in turn made price fluctuations more frequent. They also worked late to keep up with markets overseas and traded news with an ever-expanding circle of contacts. As Tom Standage writes in *The Victorian Internet,* the businessman E. W. Dodge wrote about this new way of life with a palpable weariness: "The merchant goes home after a day of hard work and excitement to a late dinner, trying amid the family circle to forget business, when he is interrupted by a telegram from London, directing, perhaps, the purchase in San Francisco of 20,000 barrels of flour, and the man must dispatch his dinner as hurriedly as possible in order to send off his message to California."

This enervation would only increase. By 1888, the telegraph had become primarily a mode of business communication in the United States. Forty million telegrams were sent that year. Just 5 percent of telegraphic revenues came from messages between family and friends; 8 percent came from news service use of the wires. The vast majority, *The New York Times* reported in a news article, 87 percent, was "commercial and speculative," a category that included not just business use of the telegraph but also wires sent to "'bucket shops' and pool rooms, where chances are sold on races thousands of miles away." For this reason, when the Royal Mail took over the British telegraph service in 1870, American businessmen were not in a hurry to urge their government to follow suit. Doing so had run the Royal Mail into debt for the first time, and taxes were raised to lower the cost of telegrams to six cents for ten words, including the addressee, or the equivalent of about a dollar today. American businessmen, then as now opposed to taxes, wondered, "Is it fit and proper," argued one of them in a newspaper story of the time, "that the whole of people should be taxed to subsidize a cheap telegraph?"

The anxiety produced by this much-increased interconnectivity trickled down, as people who thought and wrote about communication and traveled in business circles foresaw a new era descending upon them. Commentators began to worry that somehow, a more civilized time of communication was being Morse-coded out of existence. "We are in great haste to construct a magnetic telegraph from Maine to Texas," goes a famous comment by Henry David Thoreau, "but Maine and Texas, it may be, have nothing important to communicate." Progress began to take on the dimensions of Icarus's flight. "Our desire to outstrip Time has been fatal to more things than love," wrote an editorialist in the *London Star* in 1901. "We have minimized and condensed our emotions. . . .We have destroyed the memory of yesterday with the worries of tomorrow. . . .We do not feel and enjoy; we assimilate and appropriate." The writing desk of that era wasn't a place of leisurely communication but rather an indictment: "In the secret drawer the checkbook nestles comfortably close to a few brief notes and telegrams that make up the sum of modern sentiment."

Since most people didn't send and receive telegrams regularly, the telegraph made the biggest impact in their lives by increasing the scope of the world it brought to them. This new, globalized sense of now would soon test the limits of human empathy. Small-town residents in the United States suddenly found it difficult to put local news into the context of large-scale disasters around the world. One newspaper, the *Alpeno Echo* in Michigan, defiantly shut down its incoming telegraph service, tired of becoming the world's echo chamber rather than a record of its own community. "It could not tell why the telegraph company caused it to be sent a full account of a flood in Shanghai, a massacre in Calcutta, a sailor fight in Bombay, hard frosts in Siberia," Standage wrote, "and not a line about the Muskegon fire."

American Nervousness

Many dilemmas of our own age can be glimpsed in the nineteenth century's convergence of the technologies of the railroad and the telegraph with the introduction of standardized time. Information overload is beginning to create a free-floating anxiety. In *American Nervousness,* George Miller Beard discusses the impact of the increasing burden of time in the workplace. In one section, describing the modern man who has turned his watch into a fetish object, he writes:

> Before the general use of these instruments of precision in time, there was a wider margin for all appointments . . . men judged time by probabilities, by looking at the sun, and needed not, as a rule, to be nervous about the loss of a moment, and had incomparably fewer experiences wherein a delay of a few moments might destroy the hopes of a lifetime. A nervous man cannot take out his watch and look at it when time for an appointment or train is near without affecting his pulse, and the effect on that pulse, if we could but measure it, would be correlated to a loss to the nervous system. Punctuality is a greater thief of nervous force than is procrastination of time. We are under constant strain, mostly unconscious, oftentimes in sleeping as well as in waking hours, to get somewhere or do something at a definite moment.

Beard's analysis of the woes of the nervous man were part of a worldwide fad of diagnosing as neurasthenia the mental exhaustion of mostly upper-class individuals involved in sedentary employment. William James popularized the diagnosis, dubbing it "Americanitis," since Americans seemed particularly prone to the disorder due to their stressful business environment and the rapid urbanization of their society. Given how many symptoms were grouped under the rubric—everything from chronic fatigue syndrome to irritable bowel syndrome (though these afflictions were not called by those names back then)—it seems more plausible today to treat the rash of diagnoses as a visible symptom itself, the mark of a society undergoing great change.

If we simply swapped his watch for a BlackBerry, would Beard's nervous American man be recognizable today? Yes and no. In spite of the ubiquity of cellular phones, all of which keep time, watch sales have not plummeted recently; their appeal as accessories of style and status has, in fact, only increased. A watch suggests a serious, sober man even more than before. On the other hand, that nervous man on the run would have a very different sense of now from the now we live under today—or did one moment ago. It's quite possible this gentleman had a phone in his home, but after shutting his front door and stepping onto the elevated subway, he would effectively have been in a communication blackout until he reached work.

Things change even more when he arrives at the office, where telegrams would be sent and received, and there would be phone calls to take and memos to sign off on. If the man was a lawyer, he would do no typing whatsoever. Men of stature did not type—they dictated. He would also receive very few, if any, letters or notes from friends. Most of the telegrams he received would be short. None of them arrived with an attachment, unless you considered the voice component of a singing telegram, which arrived in 1933, the first one sung by Rudy Vallee, or the Candygram, which Western Union established in the 1960s.

In contrast to e-mail, all of these missives arrived by hand from a messenger boy who could have been a young man looking to get his leg up in business. As late as the 1970s, office buildings in New York City had tubular slots where a telegram messenger could come and pick up messages that had been signaled for pickup by a bell linked to a central processing station. Before their numbers were thinned out by a pneumatic tube system that carried messages to and from main hubs, Wall Street crawled with these busy runners, who snaked through the city streets and bolted up staircases. The inventor Thomas Edison, who adapted the telegraph to create the ticker tape and later invented the lightbulb, and the financier Andrew Carnegie, both got their start in life this way, meeting men of power by taking them urgent messages. A century later, the novelist Henry Miller also worked as a messenger boy, and subsequently the crisscross connections of the metropolis exploded on his pages:

> It's a human flour mill. . . . Names and dates. Fingerprints, too, if we had the time for it. So that what? So the American people may enjoy the fastest form of communication known to man, so that they may sell their wares more quickly, so that the moment you drop dead in the street your next of kin may be apprised immediately, that is to say, within an hour, unless the messenger to whom the telegram is entrusted decides to throw up the job and throw the whole batch of telegrams in the garbage can. Twenty million Christmas blanks, all wishing you a Merry Christmas and a Happy New Year, from the directors and president and vice-president of the Cosmodemonic Telegraph Company.

For Those Who Can: Keep It Short

It's worth belaboring the importance of brevity in these messages. Telegrams conveyed urgency, but very rarely did they contain large amounts of text—except those sent through diplomatic and news channels. As a result, telegrams were not used to shuttle complex thoughts from one distant location to the next, even when sent by writers. When his wife gave birth, James Joyce, who was responsible for writing some of the longest sentences since Henry James, cabled his brother SON BORN JIM. The editor Maxwell Perkins, famous for his work with Ernest Hemingway, was even briefer: GIRL, he telegrammed his mother.

Humor, which depends in part upon brevity's torque, also flourished over the telegram. Upon arriving in Venice, Robert Benchley wrote to his editor at *The New Yorker*, STREETS FULL OF WATER PLEASE ADVISE. Mark Twain once used the device to play a dirty trick upon a dozen well-known men. FLEE AT ONCE, he advised them, by telegram. ALL IS DISCOVERED. All of them left town immediately. Perhaps the shortest telegram ever sent traveled between Oscar Wilde and his publisher, from whom he wanted to know how sales of his new novel were going. The message: "?" The reply: "!"

Hostility could be conveyed, loud and clear. In 1919, the humorist James Thurber heard from his ex-girlfriend, Minette Fritts, that she had married. He cabled back directly: WHAT THE HELL! When William Faulkner submitted the manuscript of *Absalom, Absalom!,* his editor was away and the book fell into the lap of a younger editor, who wrote back criticizing the Mississippi novelist's syntax and sentence structure. Faulkner promptly responded with a telegram: WHO THE HELL ARE YOU? In 1944, Ernest Hemingway got tired of spending so much time apart from his third wife, the war reporter Martha Gellhorn: ARE YOU A WAR CORRESPONDENT OR WIFE IN MY BED? Secret lovers, no longer reduced to communicating by code in newspaper classified ads, used it to send missives—also delivered in code. Graham Greene once cabled his mistress, Catherine Walton, from Morocco in 1948: DO YOU LIKE ONION SANDWICHES GREENE, onion sandwiches being code for a sexual act.

Only the very wealthy—or institutionally underwritten—could afford the luxury of such private correspondence over the wires. Very, very few individuals had telegraph lines installed in their homes. Sending a telegram was reserved for important news: announcements of births or deaths, the latter of which were noted with a black-bordered envelope. In some cases, families received yet another telegram when the body of their loved one began its train trip home: BODY OF MAJ R ROBBINS GOES NORTH TODAY, one of them read in 1864. In other instances, an early telegram of a wound proved too hastily sent. The family of Yankee soldier Stanley Abbott received a telegram that Abbott had suffered a chest wound. DOCTOR SAYS NOT MORTAL, it said. Abbott died the next day.

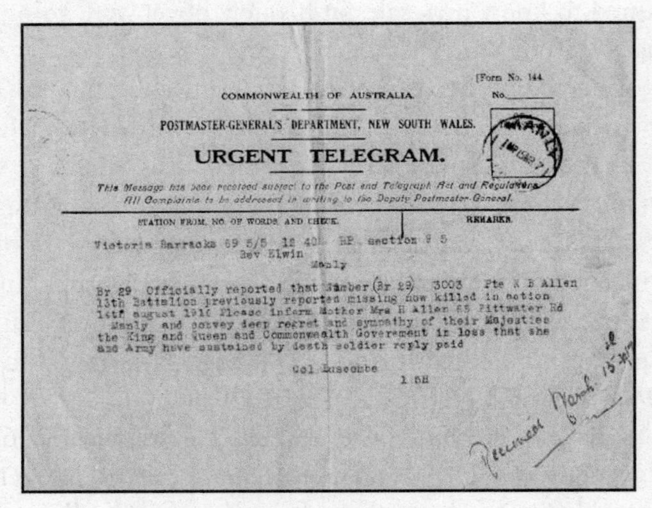

[Form No. 144

COMMONWEALTH OF AUSTRALIA.

No.

POSTMASTER-GENERAL'S DEPARTMENT, NEW SOUTH WALES.

URGENT TELEGRAM.

This Message has been received subject to the Post and Telegraph Act and Regulations. All Complaints to be addressed in writing to the Deputy Postmaster-General.

STATION FROM, NO. OF WORDS, AND CHECK. | REMARKS.

Victoria Barracks 69 5/5 12 40 BP section 9 5
Rev Elwin
Manly

Br 29 Officially reported that Number (Br 29) 3003 Pte K B Allen 13th Battalion previously reported missing now killed in action 14th august 1916 Please inform Mother Mrs H Allen 65 Pittwater Rd Manly and convey deep regret and sympathy of their Majesties the King and Queen and Commonwealth Goverement in loss that she and Army have sustained by death soldier reply paid

Col Luscombe
1 5H

Received March 15

Australian War Memorial Negative Number RC04150

The cost being what it was, even businessmen and bankers, who by 1860 relied quite heavily on the telegraph, sent just seven to ten telegrams a day. Gradually, however, prices declined and the volume of telegrams increased. In the telegram's heyday, you could send a ten-word telegram for around twenty cents and a weekend cable from New York to London for just seventy-five cents, or the equivalent of $6 and $18 today, respectively. By the end of the nineteenth century, Western Union was transmitting 58 million telegrams a year and needed to outfit 14,000 uniformed messengers. The company began offering specially designed decorative telegrams, some of which, much later, were designed by Norman Rockwell and other artists.

By 1903, the number of telegrams in all countries around the world reached 1 million per day, with Britain leading the world at 91 million a year. The United States was a close second, followed by France, Germany, Russia, Austria, Belgium, and Italy, in that order. The invention, refinement, and adoption of the telephone gradually chiseled this number down, and the ever-swinging pendulum between written and spoken language swung back toward oral communication. In the United States, there was only one long-distance line in 1885, stretching from New York to Philadelphia. By 1895, the phone system had more than 265,000 miles of wire and was growing exponentially. That year, 750 million telephone calls were made, ten for every man, woman, and child in the United States. And at the turn of the century, telegrams were outnumbered by phone calls fifty to one.

The decline was precipitous after 1945, the year 236 million messages were sent over the telegraph network. In 1950, Western Union tried to relaunch its service by calling attention to the fact that sending words instantly meant something more than hearing a voice on a phone. "A telegram commands attention—gets results" was its slogan. By 1960, though, telegram volume had dropped to half of what it was in 1945, and by 1970 it was halved again, to 69 million.

Despite their efforts, by the late 1980s telegrams made up just 5 percent of Western Union's business. The bulk of these were Opiniongrams, messages used to lobby politicians that could be sent for $5.95 for twenty words. During the Iran-contra hearings, 150,000 telegrams piled in with words of disgust and support for Oliver North, the disgraced colonel involved in shipping arms to Iran so that military aid could be funneled to contra rebels via Saudi Arabia. It was a record blip in a failing service, so the company extended its Opiniongram rates to cover messages sent to anyone testifying before a congressional investigating committee.

What Time Cannot Contain

By the middle of the twentieth century, the world had an incredible array of channels by which to keep up with one another and the news. But all this communication—all this present-tense urgency—couldn't conceal one overarching problem, one that philosophers have wrestled with incessantly. Time can be standardized, but it is not uniformly felt. As John Berger has written, "Despite clocks and the regular turning of the earth, time is experienced as passing at different rates. This impression is generally dismissed as subjective, because time, according to the nineteenth century view, is objective, incontestable, and indifferent; to its indifference there are no limits."

The friction between our private sense of time and the objectively observable notion of Time—be it in the news or in the updates of a friend talking to you by telephone—creates the texture we know as consciousness. It is the source of our subjectivity and our greatest pain and dislocation. Consciousness is what sets us apart from so many other animals and from one another. As our world moved faster, or appeared to, we became left behind and alone. A telephone call, a face-to-face meeting into which we become absorbed, a deeply rewarding mental or physical task, a moment's prayer might briefly make us forget this fact; but when the silence returns—as it must—we cannot escape the hard truth that we are alone in the world.

None of the technological inventions of the nineteenth and early twentieth centuries gave us a total reprieve from our singularity; they merely swathed it in padded wrapping. People in many countries could get into cars and drive to see one another. The telegraph meant that urgent news could be sent swiftly and business transacted. In the end, though, long-distance rates meant that it was more practical for people who had a lot to say to simply pick up the phone. One could cook or put children to bed and meanwhile carry on a separate conversation. Up until the invention of the fax machine, important memos and business documents were still sent by letter, airmail if necessary.

When it came to telling the story of one another's lives, though, or sharing the details of an afternoon's idyll, the rest of the world bridged this gap—this gap between *our* time and Time itself, from *our* minds and someone else's—as it had since literacy became widespread. People withdrew and dialed into the inner quietude, that absence of sound we think of as *our most intimate selves,* and transformed it so that they could pass it along, share it. In other words, they wrote a letter and put it into the mail. Dramatically, radically, that would change at the end of the twentieth century.

3

ALL TOGETHER NOW

The infrastructure we will need in the 21st century goes beyond traditional public works projects. . . . I envision a national computer network linking academic researchers and industry, using the nation's vast data banks as the raw material for increasing industrial productivity and creating new products.

—Al Gore, *1988*

We want to eliminate distance as a factor. . . . You can compare this to the kinds of things that happened in the 50's in the United States. We need a project of the scale of a National Highway Project for computer information.

—Robert Haber, *1988*

All the heckling radio hosts, wisecracking comedians, and savvy computer gurus were right: Al Gore did not "invent the Internet." It's also worth pointing out that the vice president never claimed to have done so; he merely brought a bill before the U.S. Senate aimed at creating a much larger, more user-friendly version of a network that already existed—ARPANET—and then made the mistake during his first presidential campaign in 1988 of telling Wolf Blitzer on CNN that he had "taken initiative in creating the Internet." Republican congressman Dick Armey pounced on Gore's apparent hubris and put out a press release mocking the Tennessee senator, and in typical say-it-often-enough-and-it's-true fashion, the word "create" evolved into "invent." The transformation was so convoluted that Stanford University researchers actually did a media study on it in which they concluded, "Truth does not always win out in the marketplace of ideas, even when the marketplace is highly competitive."

One of the reasons the story is believed to be true, though, has to do with the long gap between the development of the technology and networks that became the Internet and the public's sudden and swift gravitation online in the mid-1990s. For most people, the Internet seemed to appear out of thin air in 1995. But ARPANET, the Internet's granddaddy, was approaching thirty years of age at that point, long enough to have developed a lore and a following, not to mention a few name changes. Its first node was installed in UCLA in the late 1960s by a group of inspired, sleep-deprived researchers with a practical dream, and three more nodes were quickly added. They wanted to network with other academics around the country their hulking, washroom-sized supercomputers, machines so expensive and complicated to maintain that only a few universities in America had one.

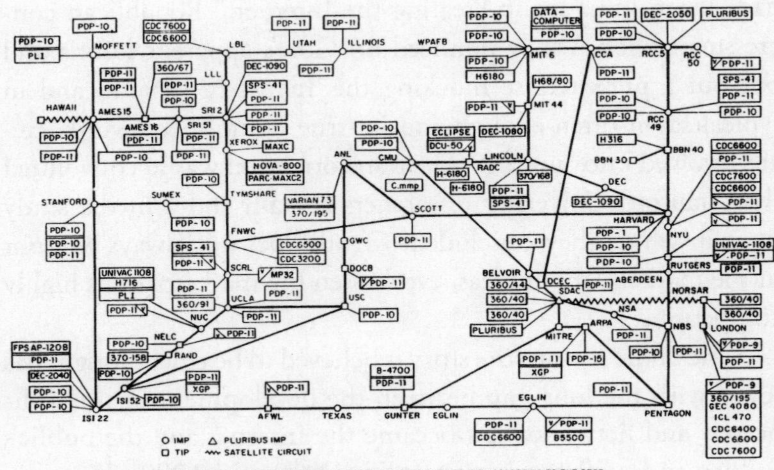

The institutions and players that turned this dream into reality echoed the overlapping forces that had brought mail to the masses and the telegram to life: scientists needed a network so they could conduct research; the U.S. military badly needed a new kind of communication network to stay one step ahead of, well, everyone else. By the late 1960s, real-time computing—hardware and software systems that work together in highly deadline-oriented situations, as in antilock brakes—had been in use in radar and missile systems for two decades. ARPANET was that technology's practical offshoot and fail-safe. A decentralized, secondary communication network would be essential in the event of a nuclear attack; given the climate of fear and dread in the United States in the 1950s, this was hardly the stuff of science fiction. Circumventing such an attack was not the explicit goal of ARPANET, but it was a prevailing preoccupation of one inventor who helped to make the whole thing possible: Paul Baran.

Like Morse's, Baran's impact as an inventor stretches across several practical realms. He is credited with inventing the airport metal detector, as well as the technology that is essential to the ATM and the DSL modem, which connects a computer to a high-speed phone line. But his place in history was secured by work he performed at the RAND Corporation between 1959 and 1965. RAND was a good home for a man worried about the future of American security, as Baran was. Founded in 1946 by the U.S. Army Air Corps, the corporation was set up to maintain the research capability built up by the United States during World War II. It was originally part of Douglas Aircraft Company, but in 1948 it was spun off and became its own independent nonprofit group. Between 1950 and 1970, Rand's growing ranks of theorists and eggheads worked on systems analysis, game theory, reconnaissance satellites, advanced computers, missile defense, and intercontinental ballistic missiles. RAND also advised Robert McNamara on the failing Vietnam War, a role that became a political controversy when a RAND employee, Daniel Ellsberg, leaked seven thousand pages of classified documents to *New York Times* reporter Neil Sheehan. They would come to be known as the Pentagon Papers.

Baran's work at RAND would never earn him the fame or notoriety that Ellsberg's act of patriotism did, but it made a larger impact on the day-to-day life of people the world round. He had two strokes of genius, both of which would refine the structure of communication networks henceforth and assist the birth and incredible growth of what became known as the Internet. He pulled them both off by overshooting expectations. At the time, the military was merely concerned with preserving "minimal essential communications," which Baran described in an interview as thus: "a euphemism for the President to be able to say 'You are authorized to fire your weapons.' Or 'hold your fire.' These are very short messages. The initial strategic concept at that time was if you can build a communications system that could survive and transmit such short messages, that is all that is needed."

After talking to generals, though, Baran realized there would be an immediate need for such a network to carry more data. When he couldn't get a clear answer for how much more, he simply decided to design something that would be able to handle an almost unlimited amount of information. His solution was to sketch out what is now called a distributed network. In a centralized network, all roads lead back to the same place; in a decentralized network, all roads lead to a number of different places. A distributed network has no center. Rather, it resembles a map of the human brain, with each "node" connected to several others. This meant that no attack on a central node could knock out all communications.

Second, and more important, Baran suggested that messages could be broken up into pieces and sent along this network, then reassembled at the point of their destination. Thanks to the advent of digital technology, data could be encoded into a series of 1s and 0s, and, if marked properly, a message could travel the most efficient route possible to its destination. If there were a slowdown in one place, the message would simply take another route. In her fabulous history of the Internet's early days, *Where the Wizards Stay Up Late,* Katie Hafner uses the metaphor of shipping a house from Boston to Los Angeles on the American interstate highways to explain why this is effective. "As long as each driver has clear instructions telling him where to deliver his load and he is told to take the fastest way he can find, chances are that all the pieces will arrive at their destination in Los Angeles, but if each piece of the house carries a label indicating the place in the overall structure, the order of arrival doesn't matter. The rebuilders can find the right parts and put them together in the right places."

Baran's ideas were eventually embraced at RAND, but, as with Morse and his telegram, the world did not leap to assist him. Five years and numerous meetings with AT&T later, in which the long-distance carrier refused to even allow RAND to test its wires, Baran gave up and moved on. As with the telegram, it turned out that Baran was not alone with his invention. Over in Britain, Donald Watts Davies of the National Physical Laboratory had been playing with a very similar idea, which he decided to call packet switching. Finally, back in the United States, a third man, Leonard Kleinrock, had written a dissertation that put together a mathematical theory of packet switching, which would prove essential to the Internet, and was developing early thoughts about how to put it into action.

Through luck and smart hiring, these men and their ideas all came to bear on a fledgling research agency started up in the Defense Department called the Advanced Research Projects Agency (ARPA), which was formed in 1958 in direct response to the Soviet launch of *Sputnik*. Within two years of its creation, though, when all space-related research was transferred to the National Aeronautics and Space Administration (NASA), also created in 1958, ARPA had nearly become redundant. The agency's response was to begin directing its energy to exploratory research programs—and after many long hours and false starts, ARPANET, the world's first operational packet-switching network, was one of its finest results. The first message was sent across the network on October 29, 1969.

29OCT69	2100	LOADED OP. PROGRAM FOR BEN BARKER BBN	CSK
	22:30	Talked to SRI Host to Host	CSK
		Left op. imp program running after sending a host dead message to imp.	CSK

This written record shows a log of the first message sent; it is not the first e-mail ever sent. Messaging between terminals linked to a mainframe computer had existed since the early 1960s, and by the late 1960s more than a thousand people used this tool at MIT alone. But sending messages from one separate computer to the next over a network was new and the beginning of a brand-new age of communication. In 1973, Ray Tomlinson would make it even easier by using the @ symbol to separate an address from the domain name, making e-mail addresses easier to write out; it was as if suddenly e-mails had a zip code system. E-mail was on its way.

As a network, though, ARPANET still had a long way to go before achieving the dream of simultaneous computing. But on one front it was an immediate and almost instant success. In 1973, a study was commissioned to see how the new network was being used. As it turned out, people weren't using it so much to share computing power, as intended. In fact, the eggheads who had access to it were doing something else entirely, not always productively, in a precursor to our situation today: 75 percent of the traffic over ARPANET was e-mail.

Plug In, Tune In, and Log On

Over the next two decades, the creation of additional networks; cheaper connectivity costs; the codification of universal mail protocol, which allowed different systems and machines to pass messages along; and the development of faster connection speeds turned e-mail from a techie toy into a household hobby. Telenet, a civilian equivalent of ARPANET, was launched in 1974 by Larry Roberts, one of the men who had shepherded ARPANET into existence, and it had enough promise that GTE bought it in 1979; Usenet, which allowed remote users to access bulletin boards, was launched by two Duke University graduate students that same year. By the early 1980s, there were dozens of networks in existence, and the number of host computers—essentially, machines connected to the Internet—most of which, by a large margin, were in North America, exploded from 213 in 1981 to 28,174 in 1987 to 313,000 in 1990. Users followed. In 1988, just 10 percent of the 19 million PCs worldwide were connected to the Internet; by 2006, Intel estimated that more than 1 billion PCs were connected, a growth made possible by rapidly decreasing PC costs and the introduction of good browsers.

The online universe that came into being from this amalgamation has radically altered human life in more ways than simply making it easier to purchase movie tickets. It altered our sense of time and space just as radically as the telegram did; ushered in an era of continuous news, since online news sites could be updated from minute to minute; and made it easier than ever to communicate at letter length at such little cost that it quickly spawned a culture of e-mail overload. Why bother sending a letter when you can just fire off an e-mail for free and have it get there instantly? More important, though, it brought people into an ever-tighter embrace with machines, which, from the beginning, were intended to blur the boundary between "inside" and "outside," a confusion that would later make it an irresistible medium for advertisers (in the form of spam).

It's worthwhile to detour for a moment to think about the PC's role in bringing this about. As John Markoff describes in *What the Dormouse Said,* the PC grew out of the countercultural environment of California in the late 1950s and early 1960s, when computer research, clinical trials of psychedelics, and the literary experiments of the Beat movement combined and overlapped in groups that bounced between San Francisco and Palo Alto to the East Bay. The goal of the PC—like that of Jack Kerouac's "The Scripture of the Golden Eternity"—was mind expansion, so much so that Douglas Engelbart, the Stanford Research Institute researcher who coinvented the mouse and did more than any designer to affect how computers present themselves to us, titled his seminal paper about the future of computing "Augmenting Human Intellect: A Conceptual Framework." The computer was not just an expensive calculator in Engelbart's vision; it should and could become an intimate, essential part of our thinking tasks.

Four decades on, Engelbart's vision of the man-computer embrace is a reality. How many of us can imagine going to work without one now? Can we even imagine working on a PC without Internet access? Not since the typewriter has a piece of machinery and what it's capable of changed work life in such a dramatic way. A typewriter helped you type faster. It also made word processing easier, especially once erase and edit functions were added, tilting the machine toward a word processor. The standardized keyboard on typewriters also introduced a new interface to writing—for the first time in the history of writing since pencils and pens became widely available.

The PC, however, introduced an entirely new way of living and doing business by becoming the portal through which *all of our work is done.* It is now a filing system, data processor, calculator, print shop, editing room, research library, music repository. In recent years, through services such as Skype, it has acted as a telephone and as a videoconferencing chat room. Once Internet browsers that could depict Web sites graphically evolved—programs such as Mosaic and Netscape—the computer became a reading library, a research room, a powerful orienting tool for our *whereness* in the world. In *Interface Culture,* Steven Johnson argues that this is such a powerful shift as to blast open an entirely new space: information space.

The mouse and the keyboard became our compasses for navigating this new realm. In the early days of the Internet, there was so little online that the computer was as necessary to modern life as, well, a compass; we use a compass now only if we're going off the beaten path, exploring. Once the Internet began to fill with electronic representations of every aspect of modern life, from furniture stores to newspapers, and those online portals created a new epistemology, a realm in which information links can be swiftly, instantly followed by the physical click of a mouse—let alone new markets, new friendships, and new ways of doing business—it was impossible to imagine life without it.

Not surprisingly, we now spend an enormous amount of time in front of our machines. Sixty-five percent of North Americans spend more time with their computer than their spouse. Far more than the dream of simultaneous computing has been achieved; this is a marriage. As Johnson points out, this symbiosis has been helped along by the artfulness of modern interface design. (Our partners are pretty!) We no longer need to punch-card code into computers or stare at a horrendous lemonade-colored blinking cursor. The point-and-click mouse design that Engelbart conceived and that was first popularized by the Macintosh, then duplicated in every computer since, bridged our old tangible world with the new virtual one. We have a "trash can" and file folders. We get mail. No longer do we have to put a picture on our physical desktop of our kids or our spouse; it can be our screen saver. This virtual desktop is our perch, our catbird seat, our platform into the world of information space, a wormhole out of it into other people's lives and to-do lists.

What this marriage lacks, however, is physical passion. Computers have become handier, cuter, some might even say sexier, but they do very little to engage us as physical beings. They have almost no smell; only the most fanatical have tried licking them. Until recently, their only sounds were blips and bleeps. Clicks of the mouse can be made with the slightest movement of one hand. Indeed, the one sense they engage overwhelmingly is sight. Our eyeballs. Light beams out of the screens and into our eyes, all day for the deskbound. We move the mouse and watch it navigate the desktop; reload Web pages and watch images appear. It is an interactive medium, which is why it's vastly superior to television, but only along one part of our physical being. The rest of our senses are effectively browned out.

And so the rise of this new way of living and working has given us a somewhat frightening twenty-first-century update to Ralph Waldo Emerson's idea that it was possible, in a heightened state of nature, to become a transparent eyeball, to let that which is the world pass through us and obliterate the subject-object dichotomy. "I am nothing," he wrote. "I see all."

Today's eyeball, however, is at the center of the world; it is pitched to, delighted, dazzled, and in control. This iris is made not for eye contact but for receiving and processing. It is—for a vast part of the day—a data processor.

The One-to-Many Rave—from 0 to 35 Trillion

Perhaps we have evolved enough in the past two centuries to enable us to live this way. I don't believe this is the case. Humans may adapt to our environment, but we also rebel against adaptations that aren't working; we mark and sabotage. E-mail has become our primary weapon in marking this rapidly expanding, radically desensualized information space with ourselves, to try to make it more human. And it is perfectly suited to the task because it allows us to spread our thoughts and words faster and farther and wider than ever before.

Until the Internet and e-mail came into being, our primary mode of communication existed on a one-to-one level. We wrote a letter to one person; we placed a phone call and were connected. At work, we could copy a memo and send it to several recipients, but even this required a further step of making copies and then physically delivering them, so the message had better be important. As recipients of media, from television to radio to newspaper and book publishing, we were locked into the other end of a one-to-many model. A broadcaster or editor decided what we might like to listen to or watch on television and sent it out over the airwaves.

The Internet and more specifically e-mail radically altered this process. By simply typing an additional address, we could send the same message to as many recipients as we had addresses. We could forward and duplicate messages as fast as the most heavy-duty copier, faster than any interoffice messenger. We have no need for a young Andrew Carnegie. The only limit to the size of our reach was the size of our address book. We could even e-mail the CEO, and chances are—until recently, when e-mail became untenable for many executives (though not all)—he or she would read it. This instantly turned anyone with an e-mail address into a broadcaster on a small scale; a large scale for so-called power users, who in the early days of e-mail were men and women who received two hundred e-mails a day.

This shift, from receiving to generating media, has created an enormous epistemological shift between reading and writing, from talking to writing. Reading, by virtue of the constant interruptions we face due to electronic communication, is harder than ever before, whereas typing and publishing have become easier than at any point in human history. Walk into any café across America, and you will witness a stirring example of this phenomenon. Whereas once cafés were filled with people talking to one another or reading books or newspapers, now you will find people sitting alone before the glowing screen of their laptop, typing e-mails, working on documents, chatting with friends a thousand miles away, or surfing the Internet. Sit down with a friend for a face-to-face chat, and you may be scowled at.

In *On Photography,* Susan Sontag wrote about the way that the domestication of photographic technology allowed people to believe that all the world's images could be indexed, possessed, known. The explosion of e-mail and other text-based communications and the phenomenal ease with which these technologies can be used has done a similar thing to words. If we know something, experience it, see it or do it, complete a task at work, we must record this fact in type and share it with another person. We may not be able to index the known world, but we can create a word-based library of ourselves from moment to moment if we type fast enough and keep in touch with enough people.

Universities, which are stocked with young people eager to chronicle their daily revolutions, were a hothouse for this growth, and it demonstrated how connectivity expands exponentially in an inbox. The explosion of its use on the campus of Simon Fraser University in Canada makes for an interesting test case. SFU was one of the first universities to offer e-mail service to its faculty in 1983, and over the next few years it joined various external networks. In 1986, fifteen hundred students and faculty were using the service, generating between ten and twenty thousand messages per month. Five years later, the number of messages per month had risen to seventy thousand. By 2006, the number of e-mail accounts had increased to forty thousand, but the number of messages had blasted off into an entirely new galaxy. Each month, SFU users were generating 10 million messages—a *14,000 percent* increase.

The exponential increase seen at Simon Fraser was witnessed across the Internet and around the world. In 1992, just 2 percent of the U.S. population used e-mail; that jumped to 15 percent in 1997, and by 2001 it was estimated to have leapt to 50 percent, perhaps due largely to the number of people who were wired at work. Early stories about e-mail recall Laurin Zilliacus's breathless wonder at the journey his letter made from Lapland to the American heartland:

> John M. Woram went to his mailbox in Rockville Centre, L.I., recently and mailed a letter to a colleague in the Galápagos Islands, 650 miles off the coast of Ecuador. His letter arrived there in five seconds. A reply was waiting in his mailbox the next morning.
>
> "My, how the mail has evolved," said Mr. Woram, who is writing a history of the islands made famous by Charles Darwin. "It used to take as long as seven months to get a letter there and back."
>
> Mr. Woram's magical new mailbox is inside his personal computer at his home, and his correspondence with the Galápagos now travels at the speed of electricity over the global computer network known as the Internet.

This was in the middle 1990s, the era of the Internet's explosion, when the U.S. Postal Service's total number of deliveries was ten times the number of electronic messages worldwide, when just 5 percent of American households had a modem. That would change quickly, as many people discovered how easy and convenient it was to send and receive electronic mail. Not surprisingly, corporations and offices embraced e-mail. Between 2000 and 2002, the number of workers with access to e-mail on the job in the United States almost doubled from 30 million to over 57 million, 98 percent of whom had access to their own account. By 2000, there were 4 trillion e-mails sent globally. In 2007, that number hit 35 trillion—a number that dwarfs the number of text messages (3 trillion) and telephone calls (165 billion minutes in 2005).

Two trends really helped domesticate e-mail. First, it became widely, easily, cheaply available at home through dial-up service, which had indeed existed for more than ten years, but mostly for Bulletin Board Systems (BBSs) and owners of specific machines. It truly became an industry in 1995, the year that Prodigy, CompuServe, and especially AOL began offering it over the Internet to customers who were using any kind of PC. In 1995, AOL had 5 million subscribers. By 2003, it was up to 35 million.

The invention of handheld devices played a role in pushing these numbers even higher. In the beginning, these tiny machines were merely very small (and very expensive and easily lost) address books. In 1993, Apple introduced the Newton, which it marketed as the ultimate information tool. In terms of communication, people who wanted to communicate wirelessly were far more likely to own a pager. In 1980, when pagers had a short range of communication, there were 3.2 million in use worldwide. As their range expanded, so did their use: by 1994, there were more than 62 million pagers in use around the globe.

The first BlackBerry was introduced in 1999, and though it ran on AA batteries, it could handle e-mail, paging, and a few organizing functions. As the machines improved, their use skyrocketed. Between 2004 and 2008, the percentage of people checking their e-mail on a handheld doubled. In one survey, 59 percent of the people who used such devices read their e-mail as soon as it arrived.

The Virus of Consciousness

Aside from the obvious reasons for this explosion—e-mail is cheap, fast, easy to use, and a lot of people are reachable through it—there is another explanation for this behavior, one that reaches back to the roots of the personal computer and the mind-expansion goals of its creators. The machine did, in fact, become a virtual extension of our minds—an orienting tool, an organizing tool, a tabula rasa upon which we can express our thoughts, and a computing tool all at once. And by virtually extending this surrogate brain into the Internet, we became linked with all the other mind spaces of people who were linked in. In fact, one of the most popular social networking sites on the Internet is called LinkedIn, which allows people to create an online address book of their professional contacts.

As Clay Shirky has astutely noted, this network of connectivity has created new kinds of group collaborations, from Wikipedia to the loose, fanatical collaboration of programmers who work—for free—on the Linux operating system code to make it the best in the world. But it's also created a physical, external, ever-present viral port for a part of our life that until now has either remained private or moved at a much slower rate: our consciousness. "We are all tainted with viral origins," says one character in a William S. Burroughs story. "The whole quality of human consciousness, as expressed in male and female, is basically a virus mechanism."

In other words, the Internet is the perfect host for spreading our viral minds. In the past, human consciousness could be recorded in words and spread as fast as those words could travel—which, as we've seen, even with the advent of the telegram, moved fairly slowly. The post buffered us from the full cacophony of human consciousness as it was experienced by all the people we knew around the world. With the advent of e-mail, however, the delivery of those dispatches from the minds of others began to occur more or less instantly. Moreover, they shot right into the working nerve centers of our new surrogate brains.

This is why turning on your computer in the morning—or checking your BlackBerry from home while having your coffee—can feel so overwhelming, so in-your-face, especially when it's e-mail from the office. It's not just the number of messages but the feeling that someone has *invaded your head.* So we push back—respond and delete—however, e-mail is insidious, hard to break; it's hard to figure out when a conversation is concluded. The time it requires to do this work, to type all these messages, did not come out of nowhere. In 2006, the California-based technology research firm Radicati Group reported that the average office worker sent and received 126 e-mails per day. If it took this busy bee only a minute to read and respond to each and every message, simply keeping up with correspondence would take up a quarter of his day. At the current rate of growth, the Radicati Group estimated that in 2009 the average office worker will spend 41 percent of his or her day reading and responding to e-mails.

Life on the E-mail Treadmill

The success of the PC as an extension of our brains and our working selves has been solidified by the ubiquity of e-mail. For many people, opening their e-mail program marks the start of their day. Depending on what comes in, it sets the morning's priorities and becomes a kind of rolling to-do list, which is why keeping up with it feels so important. You need to clear what comes in before it buries the list of lingering e-mails—the ones that need or require more involved responses—in a pile of electronic silt.

E-mail is asynchronous technology, meaning that two people don't have to be present for it to work. We send e-mail; it is received and eventually answered, usually the same day. Indeed, one survey of companies found that half of all respondents expected to get back to e-mail queries within four hours. The senders, on the other hand, had a different time frame for how soon they expected to hear back from an e-mail they fired off: they expected to hear back in a day. But as the volume of e-mail increased exponentially, response time decreased: more and more people found that staying atop that ever-lengthening queue kept a clear horizon inside their computer. A quick, easy question got a faster answer since it was something done, checked off, accomplished.

The acceleration of response times, however, began to train people to expect more immediate replies. Everyone has received an e-mail from an impatient coworker or friend asking, two hours later, "Did you get that e-mail I sent you?" Aside from being annoying, such follow-ups reiterate a truth that working over e-mail has made visibly clear in a way never before in the work environment: we're all connected.

If someone e-mails you asking for a piece of information and you don't reply until the end of the day, chances are you prevented a piece of work from being completed. Not all e-mailers experience this. But for the average worker, the worker who in 2009 will be sending and receiving two hundred e-mails a day, this is a burden to bear. When you arrive at work and there are twenty e-mails in your inbox, the weight of that queue is clear: *everyone is waiting for you.*

So you clear and clear and clear, only to learn that the faster you reply, the faster the replies come boomeranging back to you—thanks, follow-ups, additional requests, and that one-line sinker, "How are you doing these days?" It shouldn't be such a burden to be asked your state of mind. In the workplace, however, where the sheer volume of correspondence can feel as if it has been designed on high to enforce a kind of task-oriented tunnel vision, such a question is either a trapdoor or an escape hatch. From it you can spy sunlight, a tiny rebellion from the drudgery. So even, in some cases, if you're overwhelmed and you haven't the time to address what in any other circumstances would be a welcome interrogative, you reply. Now you're chatting. The speed at which e-mail arrives and is returned has begun to reflect early chat room discourse.

Except that e-mail can be so many different things. In some instances, e-mail can be written as properly as a legal letter. In other cases, it can be informal, misspelled, full of puns, jokes, and emoticons. Each correspondent we have, and each interaction with that correspondent, demands a slightly different register. Some days, you avoid the dangling personal detail and plunge onward. On other days, you pause and open up the door no one really wants to walk through as a make-nice gesture: "How are you doing these days?"

Juggling the flow of messages and the various response styles and registers makes the workday an exercise in linguistic multitasking. If you get a large volume of e-mail, as more and more people do, it's exhausting—trying to represent or translate all your different moods, tones, personalities, and styles into some sort of textual equivalent. And that's just the sending, where as a correspondent you have some degree of control over the opening gambit in a conversation, a pretty clear notion of what the message is intended to convey. Replying is even trickier. "When you see an Internet utterance, you often don't know how to take it," David Crystal writes, "because you don't know what set of conversational principles it is obeying." Novelists work their whole life to be able to capture this variety of their characters' modes and emotional tones in prose, and now at the rate we are typing messages, many of us are pumping out a hefty epic once a month.

The truth is that text rarely, if ever, can equal the richness of a face-to-face conversation. It's static, disembodied. It does not convey hand gestures, verbal tone, inflection, or facial expressions, things we are taught from birth to encode and decode. Indeed, these are some of the first things children learn when speaking; even before they can form words, they mimic the cadences and tones of speech they have heard. They gesture. We learn to communicate with our bodies. Talk on the phone with a friend across the globe, and you will discover how hard a habit this is to break. There are your hands, waving and punctuating, even though your friend cannot see them. It's hardwired into us. Even the blind talk with their hands.

And this is just the body. Text, until this point, has also had a kind of body: the page. Formal letters came on heavy paper stock; notes dashed off while traveling came on the stationery of hotels; postcards forced correspondents to become prose poets to close the gap between what was on the card and what they were writing; telegrams with a black rim were to be dreaded. Electronic messages are completely devoid of this sensuality; they all arrive in the same format. They have no messenger bringing them there, as an interpreter. As a result, the tone of e-mails is misunderstood more than half of the time, compared to just a quarter of the time over the phone and even less often face-to-face. A Cisco study found that workers who had to collaborate by e-mail for an extended period invariably experienced a breakdown in communication.

Coupled with the sensory brownout that the computer provides and the speed at which e-mail arrives, this means that conversations that go on too long—especially complex, detailed conversations—slide down a slippery slope toward some sort of misunderstanding. Just 6 percent of people in one survey found e-mail an effective tool for bringing up issues with supervisors, and only 4 percent found it helpful in talking about sensitive issues.

For this reason it's hard to complain about emoticons—at the rate at which e-mail comes in, they perform a necessary function by becoming what the linguist Naomi S. Baron has called a visual language. When a conversation starts veering off topic or into cloudy territory, emoticons hammer it back into shape and ham-fistedly make amends. By aping childishness, they bring us back to basic lessons, such as "Play well with others," and then boot us back into action. Most of all, they keep it simple. It's hard, after all, to misinterpret a smiley face.

Yet the misinterpretations persist and fester, especially if you don't respond to an e-mail—or if someone never gets back to you. When everyone appears to be replying at warp speed—they must be if you're getting so much e-mail, right?—the lone silent correspondent is a mysterious siren call. It encourages self-doubt. What was it you did to earn the cold shoulder? Maybe you were never actually worth the time and they just started to heed their instincts. Part of you wants to check in: Maybe the message did get lost? Another part of you doesn't want to be that person chiming in pathetically from the sideline, "Did you get that e-mail I sent?"

Meanwhile, on flows the e-mail down the screen, like a current with riptides and swirls. Enter it at the start of the day, and you're off. You paddle frantically and seem to get nowhere. Checking e-mail on vacation, at night, in the car, at the bus stop, or in the grocery store, as handheld devices have made possible in recent years—and as many people do—is like trying to stick a finger in a dam. The flow just finds a new crack, a new fissure, and before long it's pouring out again. As Mark Twain noted of his river, "The Mississippi will always have its own way; no engineering skill can persuade it to do otherwise."

Still we try to control it. We set out-of-office autoreplies when we leave an hour early for a doctor's appointment, and try to get a jump on it by checking it on e-mail-free Fridays, the day some companies have designated for workers to catch a breather from the flow. We sneak a peek before going to work and clock in before going to bed. It's our midnight snack, our reminder we are needed, the mother of all time killers. Thanks to the speed of our replies, the variety of interfaces through which we now receive e-mail, and the innumerable locales in which we check it—*can't type on this freign keybored*—it's hardly a wonder we mess up, step over the line, send it to a wrong address. And whether it's our mistake or theirs, someone will hear about it—since e-mail has gone viral.

Pay It Forward

With e-mail it's now become easier than ever to share information with other people. We just forward it. If the Swedish man living in Idaho in the last chapter were living today, he could simply send his brother a link to the newspaper article about relations between his home country and Norway. If Howard Moss had received Benchley's telegram from Venice, he'd probably forward it to someone within the office: "Look at what ole B is up to." The amount we can share, in an instant, has exploded.

There are enormous practical benefits to this capacity. We don't have to pick up the phone and give directions; we can just send out a link to a MapQuest page and let the recipient follow the route from there. If someone has been left off a party invite, we can simply forward it to them. Learn something new about a disease that afflicts a family member or get some news about a loved one traveling abroad? We no longer have to wait to pass the information along: we just forward the e-mail or attach the link to a blog post. This capacity has been an enormous boon for groups. As the author, educator, and environmentalist Bill McKibben points out, in spite of e-mail's ability to destroy productivity, it's hard to completely discount the tool: "We organized 2,000 demonstrations across the nation last year about climate change and couldn't have done so without e-mail. . . . [but] it's profoundly immature technology at this point, and we better figure out how the hell to use it without wrecking our lives or we will lose a potentially fascinating medium."

McKibben is right; e-mail is a great combination of social tool and broadcast tool. If Martin Luther were alive today, he possibly would be e-mailing his theses around instead of nailing them to church doors. But the question remains whether his ideas would be lost in the wash. The forward capacity—which bestows upon individuals a fast, cheap internal Xerox machine—means that information of all sorts travels faster than ever. And we are now more in control than ever of filtering it for one another. One of the most popular features on *The New York Times'* Web site is its "E-mail this" function, which takes the social aspect of newspaper reading and puts it on steroids. Many blogs, which are now read by more than half of all online users, have e-mail tabs at the bottom, allowing opinions and musing and responses to news and information to rapidly reach people who might otherwise never have seen them.

This culture of forwarding stuff—of sharing everything from the esoteric to the essential, or the esoterically essential—contributes to the problem of e-mail overload, since people are far more likely to e-mail a news link than read it out over the phone or print it out and mail it to a friend. The inbox quickly became our primary receiving port for pass-alongs, be they news stories or jokes—if we are at work, they will run shoulder to shoulder with other information that has been forwarded because it is presumably of interest for doing our jobs.

Forwarding—and cc-ing—is an easy way to keep someone informed of developments at work. In a culture of frequent forwarding, though, this has quickly become a mixed blessing. If we want to get the essential mail, we have to do some sifting. When everything can be forwarded quickly and for free, work that wasn't collaborative before becomes hard to conceive of doing without linking someone else in, forwarding information to a colleague so that he or she can be *in the loop*. During his 2008 presidential campaign, Barack Obama received *two* daily e-mails from his foreign policy advisers: one summarized the past twenty-four hours of events on the world stage; the other included a series of questions he might field from the media—followed by some suggested answers. These two e-mails were culled from the thoughts and input of a team of three hundred advisers. Imagine if they had all e-mailed him at once, as happens in many organizations. When not regulated, this instinct can lead to a tsunami of useless information. Not surprisingly, shortly after he was elected, President Obama got a secret e-mail address that only a handful of people in Washington know.

No one understands the need to control input more than people at the top of decision queues, such as managers and executives. When most people were loping along with ten to twenty e-mails a day a few years ago, they were already drowning in messages, a barrage that inspired a mild form of hysteria. In 2005, Sun Microsystems CEO Scott McNealy painted the portrait of a modern-day executive that sounds more like a telegraph operator: "get hundreds a day, I review them all, answer many, forward many for response, hate the junk, I type really fast, ignore perfect grammer and typin, getting more over time, but can read e-mail from any browser, have T1's into my homes and read all that is left from the day before I go to sleep after the boys go down and get up before they do to read what came in while sleeping." It's hard not to applaud McNealy for his desire to stay informed, but at what point does e-mail undo the job of delegating? Prior to e-mail, work didn't always have to be observed to get done; now it's hard to imagine completing a task without having someone virtually watching over our shoulders.

Aside from the torrents of e-mail it creates, there's an even more problematic tendency that the culture of forwarding unleashes over e-mail: it makes it hard to know when to stop. In many computer interfaces today, all e-mail essentially looks the same. For a significant portion of people, it all lands in the same box and in just a handful of scripts, as opposed to the millions upon millions of unique handwriting styles. There are no identifying, personal factors, save the name of the sender. This lack of cues can be confusing. If work e-mail, information, invitations, inspiring pep talks, and jokes from your dirty sister can all be forwarded, what can't? In this fashion, one of e-mail's primary functions has been to explode one of the overarching notions behind so much written communication up until this point, even if governments didn't always respect it: that it was private.

"Dear Asshole" and Other Errors

All communication technologies alter manners and create (and reflect) new standards as more and more people use them. Recall how the rise of the U.S. mail caused a rash of anxiety over the state of the letter—and the need to hew to proper forms of it—or how the typewriter was viewed early on as a brutish machine. In the beginning, it was considered inappropriate to send a typewritten letter for anything other than business. Before long, however, that standard had changed radically. E-mail has caused its own ripples in many spheres, from language to parenting, but it has been most fascinating when it comes to the issue of privacy.

The number of users and the ease of forwarding messages unleashed something that had previously been tamed. It is an obvious breach of etiquette to hand a letter over to anyone but its recipient; tampering with the U.S. mail is also a felony. But e-mail is constantly, continuously, recklessly forwarded. It's so easy. How many times have you received a forwarded message—sent so you have one piece of information—only to find below it an entire, supposedly private conversation thread left intact? And now yours to peruse, should you so desire.

That's the benign explanation: carelessness. But every day we encounter situations where e-mail's viral nature is clearly being exploited for payback, revenge, comeuppance. In 2006, a lawyer at a Canadian firm who felt he was being harassed for being "a snappy dresser and too friendly and too gay" didn't approach human resources but e-mailed five hundred people company-wide, writing in depth about his problems with one individual while informing his army of recipients he was "an expert yogi" and that he "trained with enlightened masters in meditation and I know a plethora of great techniques for clearing emotional poisons." The attorney knew exactly what he was doing. "I thought I'd shoot it down the associate gossip channels, just for kicks," he wrote in one message, asking the recipient to forward his diatribe. "I thought I'd mirror the model of gossip at my firm. So please indulge me in passing it along."

I didn't have to dig to turn up this e-mail. It's posted on the Internet, where you can find all sorts of supposedly private e-mails, such as the one *New York Times* op-ed editor David Shipley sent to the McCain campaign saying a piece McCain had written could not be accepted as currently submitted. For a time, all the deposed e-mails that had been sent at the Enron Corporation could be viewed on a Web site, where personal e-mails about affairs and events at home were mashed up with the e-mails that showed that employees knew what was going on at the energy giant. In 2004, *Wall Street Journal* Mideast reporter Farnaz Fassihi sent an e-mail to her friends about the deteriorating security situation in Iraq, which at the time was being played down by the Bush administration—how impossible it made her job as a reporter, let alone her existence as a human being. It quickly became a global chain e-mail. "I wrote it as a private e-mail to my friends as I often do about once a month," she wrote in another e-mail to the journalism blogger Jim Romenesko, "writing them about my impressions of Iraq, my personal opinons [*sic*] and my life here. and then it got forwarded around as you can see in a very unexpected way."

The gap between what one person and another considers private means that almost any message can be forwarded—which can lead to devastating personal experiences. In England in 2000, a woman sent an off-color joke about oral sex to some friends and received a reply from her boyfriend, which led to a humorous but intimate exchange about her performance in a sex act. The boyfriend then forwarded the compliment—and its queue—to six of his mates with a note: "Now THAT's a nice compliment from a lass, isn't it?" It was immediately forwarded to several other men, who looked up her picture on the Internet, started a campaign to find her and give her a medal, and on it went. Eventually, after the message had made its way around the world, it came back to the woman, who wrote all of the peepers—that's how she viewed them—a stinging scold:

> I Can't believe this!!!
>
> First of all, I don't know any of you!! What do you care about my social life? Don't you sad bastards have anything better to do with your time?
>
> Shouldn't you all be working. I'm going to make it a point to send this email to info@yourcompany.com just so that people can see what you do with your time!

To you Girls: You've all swallowed at one time or another, so don't judge me!

To the Guys: All you're going to get is a fantasy, so go do what you're good at . . . tossers!

Yum!

—Claire

Speed—and the ease with which e-mail can be sent accidentally, thanks to the auto fill-in-address function, to a large group—is part of the problem. In 2006, Lycos reported that every second, forty-two e-mails were misdirected. Sixty percent of these blunders involved an e-mail being sent to the wrong address, and a third of them were "steamy." Always it's information considered private pulled out of context. One executive once sent details of his salary to the entire company by accident and pulled the fire alarm in panic. A schoolgirl in Devon once wound up getting top-secret e-mails on a Pentagon round-robin list when a navy commander accidentally added her address. A professional spammer—who used to put "Asshole" into the address function of a test message—coded a message incorrectly, so that a couple million e-mails went out addressed "Dear Asshole."

It's hard not to laugh about such gaffes. In some cases, e-mail seems to out bad behavior and bring it to justice—or at least give it a good smack. Neil Patterson, the CEO of Cerner Corporation, a Kansas City–based health care information technology firm, sent a blistering e-mail to his managers when he felt that their work ethic had slacked. "The parking lot is sparsely used at 8 A.M.; likewise at 5 P.M.," he wrote. "As managers—you either do not know what your EMPLOYEES are doing; or YOU do not CARE. . . . In either case, you have a problem and you will fix it or I will replace you." The message was quickly forwarded around and wound up on Yahoo!'s message board system. The news hit Wall Street, where people assumed the company was in trouble, and its stock price plummeted 29 percent.

Politicians have proven reliably good at making e-mail blunders. On September 11, 2001, as the World Trade Center towers were collapsing, British transport secretary Stephen Byers's adviser Jo Moore e-mailed colleagues suggesting it would be a good time to bury bad news—everyone's attention would understandably be elsewhere. When the e-mail was forwarded around, it came out that she had made a similar suggestion on the day of Princess Margaret's funeral. After British prime minister Gordon Brown returned from a visit to China, a junior treasury clerical officer, Robbie Browse, sent an e-mail to friends that included racist comments about Chinese eyes. He accidentally also sent the message to eighty-six members of the press, one of whom replied, "Will we all be invited to your leaving party?"

Had such a blunder happened in a conversation, the story would be passed around, but chances are it would be changed and eventually lose momentum. In an e-mail environment, though, such a faux pas is just one click away, and an exact record of what was typed is created. The ease of duplicating it and sending it along means that an online humiliation can be experienced by vast numbers of people. The laugh track is just waiting for us. But that's not the biggest danger created by e-mail. The biggest problem has been there from the very beginning.

Door's Open, Key's in the Car

As the number of Internet users slowly increased in the 1970s and '80s, the new network became too good a target not to be toyed with by hackers—they were the twentieth-century version of the highway robbers who plagued the early mail. But there was a category difference: if mail was robbed in Philadelphia, it didn't affect a pouch going to the Ohio Territory. With the Internet, however, the connectedness was both its strength and its weakness; plant a strong enough virus anywhere in the system, and it had the potential to take the whole network down.

The rise of the Internet saw the rise of a new kind of crime, the ramifications of which would only grow as more and more systems, such as telephone switchboards, were computerized and hooked into the Web. Between 1978 and 1983, there were at least thirty attacks on computer facilities in Europe, with some of them being literally blown up. Tampering with computers was a powerful avenue for revenge for disgruntled ex-employees, especially since so many companies were so vulnerable. Two former programmers for Collins Food International were caught planting a logic bomb—a piece of code secreted in software that sets off a malicious function when certain conditions are met—in the computers that controlled the payroll and inventories of four hundred Kentucky Fried Chicken and Sizzler Steak House franchises. If it hadn't been found in time, the bomb would have erased all the records, as well as any traces of the bomb that was planted. By August 1983, $300 million a year was being lost in the United States due to the fraud and viruses perpetuated by computer criminals.

And it was easy. One description in a *New York Times* article from the period conveys the aura of mystery and criminal intrigue that surrounded the early Internet: "Penetration into the misty realm of computer networks can be easily and legally achieved by anyone with a home computer and the proper modem, a device selling for $100 or so that converts a computer's digital pulses into electromagnetic waves that can be transmitted over a phone line. One simply dials the seven-digit local telephone number of a data network and starts roaming the electronic ether." There were no firewalls, and in some places a hacker could try as many passwords as he liked. "It's like leaving the keys in the ignition of an unlocked car," said Martin Hellman, the Stanford professor credited with inventing public key cryptology.

Given that military installations and research universities were among the first institutions to become networked, there were serious security risks at the heart of the Internet from the beginning. In spite of $100 million spent to prevent it from happening, ARPANET was broken into numerous times, and in the early 1980s a motivated hacker broke through the security system of the Lawrence Livermore National Laboratory in Berkeley, California, where nuclear warheads and other weapons are designed, while another used Telenet to dip into the Los Alamos National Laboratory in New Mexico, where the hydrogen bomb and other nuclear variants were hatched. The fear of such events quickly entered the mainstream culture. In the 1983 film *War Games,* Matthew Broderick plays a young computer hacker who discovers a back door into systems just like this and winds up playing an escalating war simulation with a supercomputer at a top-secret air force installation. Due to these security breaches, ARPANET split into two networks—MILNET, for the U.S. military and Defense Department, and DARPAnet, which remained in public use for universities.

These breaches did more than just split up a few existing networks and motivate security concerns, however; they challenged the utopian dreams of the Internet. It's important to remember that, however essential Baran's apocalyptic fail-safe was, the basic thrust of the early network was an idea of sharing and collectivity. The Internet was a true tabula rasa, since, unlike the American territory, it hadn't been stolen from anyone: it had been created out of thin air. And companies followed suit with the communal spirit. "At the beginning, companies make it easy to get on and assume people are going to be nice," said the vice president for development and engineering of BBN Communications Corporation, a company that helped pioneer the development of data networks. "That lasts for a while and then you have to add access control. You can't just leave all the doors open."

No attack caused people to question this attitude more than the Morris worm of 1988, the first computer worm released onto the Internet. The architect of the worm was Robert Tappan Morris, a twenty-two-year-old graduate student at Cornell University. According to Morris, he created the program simply to gauge the size of the Internet. And he did so by exploiting a hole in the Unix operating system's "sendmail" program. The worm disguised itself as a user and sent itself out over e-mail, where, upon receipt, it would be duplicated and sent out again to a whole new address book, and upon receipt there it would duplicate again, leapfrogging off yet more address books. It was ingenious and incredibly destructive: it clogged thousands of machines and nearly shut down the entire Internet. That Morris was the son of a National Security Agency computer security analyst added a special irony to the case.

The nation's response was swift and alarmed. By the mid- to late 1980s, America's data networks were accessible by almost 9 million desktop computers. How many more Robert Morris types were lurking out there? *The New York Times* ran *eleven* stories on the story in the three days after it broke, including an op-ed by a Cornell graduate student who announced that the Internet's age of innocence had officially come to an end.

"Many of us know how to abuse the system, read others' files and steal secrets, but old-fashioned etiquette stops us," argued Peter Wayner. "The rest of us are caught in a similar bind. Do we encourage trust and freedom in experimentation or do we install complex safeguards? At America's universities, computer scientists surely must realize we can't keep leading an Eden-like simple life in the heart of the computer age."

As he himself suggested at the time, Morris did the developing Internet an expensive favor: he exposed its security weakness and planted a large reminder that what we send over the Internet is not private and that the very machine and interface through which we access information is vulnerable, too. Someone—or something—can reach down into our virtual desktop and rifle through our things, our thoughts, our financial data. We are connected, but the size of the connection is far too big to rely upon basic human trust and existing laws to protect us from malfeasance. Our new communication tool of e-mail was a boon—who could say no to receiving messages from friends and loved ones around the world?—but it was viral, and that was perfectly suited to another group who wanted something from us: advertisers.

Don't Talk to a Stranger on the Internet

Spam is such a universal problem today that its dimensions are hard to properly comprehend. By some estimates, 85 to 95 percent of all e-mail sent is spam, and dealing with it cost $140 billion in 2008. It has been with us since the beginning of the Internet, too. Gary Thuerk sent the first piece of it in May 1978 over the ARPANET to 400 of the 2,600 people who had e-mail addresses at that time to invite them to an open house for new models of Digital Equipment Systems computers in Los Angeles. Like G. S. Smith's band of circular mailers, Thuerk had to type every e-mail address in by hand. Many of the people who heard from him weren't happy about being pitched. Someone from the RAND Corporation wrote to him to say he had broken an unwritten rule of the ARPANET that it wasn't to be used for selling things. A major phoned Thuerk's boss and asked that he never send such an e-mail again. Even so, it was a cost saver and a success. It also led to an estimated $12 million in sales.

The origin of the word "spam," as identifying unwanted mass messaging, is in dispute. One of them links back to the Monty Python skit from the 1970s in which a man and a woman (played by Eric Idle and Terry Jones, in drag) are trying to order from a breakfast menu at a cafeteria in which every item has Spam in it. Spam—the canned pink pork product—was one of the only meats not subject to rationing in post–World War II Britain, so it was ubiquitous, some would say unfortunately so. Whenever the word "Spam" is uttered in the skit—and it is said 132 times in three minutes—a chorus of Vikings chimes in.

As in postwar Britain, people didn't want any spam, but they would get it nonetheless. Aside from Gary Thuerk's message, other examples of early mass messaging included one sent on an early time-sharing network mail program at MIT to the more than one thousand users linked to it protesting the Vietnam War. The message began: THERE IS NO WAY TO PEACE. PEACE IS THE WAY. In the early days of the Internet, "spamming" referred to the habit of flooding chat rooms and bulletin boards with unwanted text. Around this time the immigration lawyers Laurence A. Canter and Martha S. Siegel paid a Phoenix programmer to flood Usenet's various message boards with an advertisement for their service of enrolling people in the green card lottery. As with Thuerk's e-mail, the outcry was immediate. But it didn't matter; the scheme worked. In just two months the ad brought the couple $100,000 of new business.

As more and more people began using e-mail, spammers gravitated to it as the best way to target potential customers. By 2005, there were 30 billion spam messages *per day*; in 2007 that number had jumped to 100 billion. The number of these e-mails that are trying to sell products has also led to spam being called junk mail, a phrase that refers to the load of "junk" advertising circulars marketed to people through the post. One of the most common ways of sending messages—and eluding authorities—is for spammers to take over a series of computers, which are turned into "zombies" that work together in networks known as "botnets," and use them to send spam.

A botnet turns a series of hijacked computers, most of which are in homes, into a spam factory. Most computers become part of a botnet because they have inadequate firewall protection. A Trojan horse, or piece of malicious code, can be sent down an open line and activated later, causing the botnet to transmit messages either to a single site, shutting it down as a form of attack, or to many addresses, in the form of spam. Eighty percent of the spam sent in 2006 was sent from zombie PCs. In 2008, there were more than 10 million zombie PCs in use at any one time. In many cases, the owners of the PC never know that they have been taken over. It happens in seconds. In 2005, as a test, the BBC set up an unprotected PC, and within eight seconds it was infected by a spammer's worm.

Staying ahead of these armies requires a lot of work and

money. In November 2008, a San Jose, California, Web-hosting company called McColo was pulled offline when security experts approached the companies that managed McColo's connection to the larger Internet, showing that McColo's Web sites were being used for spamming and other online schemes. Indeed, it was estimated that 75 percent of spam shot out into the world had come from machines hosted by McColo. But the fix was short-lived. The machines, which had been infected by a Trojan horse virus called Srizbi, formed what may have been the largest botnet in the world. At some 450,000 machines, it was capable of sending 60 billion e-mails a day hawking everything from watches to penile enhancement pills. Deprived of the McColo-hosted Web sites, however, these machines lacked a connection to centralized instructions. Once the sites went down, they simply started looking for new domain names where they could find new instructions.

One security firm, FireEye, found that if it registered domain names that the Srizbi-infected computers would look for, it could actually stay ahead of the spam problem. Each week, it registered 450 new domain names at a total cost of $4,000, the idea being that it could possibly send instructions so complicated that they would halt the compromised computers in their tracks as they tried to work them out or actually send instructions to the computer to uninstall the virus program. The latter idea, however, could have been illegal or actually harmful to the infected computers. So eventually, after unsuccessful attempts to enlist other corporations, such as Microsoft, or the U.S. government to enlist the remaining domain names sought by the Srizbi-infected computers, FireEye stopped the practice. A few days later the massive botnet was resurrected and the spam volume shot up again.

Donald Trump Wants You, Please Respond

Not all scams and spam problems have passive victims. In other cases, "phishing" schemes encourage people to hand over passwords or private information by posing as e-mail from a legitimate, trustworthy source, such as a bank, a health care organization, or even the IRS. Some of the earliest cases of phishing occurred over America Online, the world's largest ISP network in the 1990s. Hackers would break into the AOL staff area and send instant messages to users currently online, posing as staff members needing to confirm password information. Even though AOL had a message on its screen—"AOL will never ask you for your password"—the scheme worked, allowing phishers to then use those accounts for spam or other malicious purposes. Breaking into an AOL internal account gave a phisher access to AOL's membership search engine, which gave access to credit cards. We've come a long way from the British-American Claim Agency's twelve typists sending out phony pitches for inheritance claims.

Since then phishing schemes have become incredibly more sophisticated and hard to stop. In recent years they have become clever enough that criminals can figure out which bank a victim might use, which is called "spear phishing." A message will be sent with a phrase such as, "We are changing systems and we need you to confirm your password data." If you click on a hyperlink in the e-mail, it takes you to a site that is bogus but cosmetically similar to that of the institution, which collects your data. We care about our money and our health, and we care about love—and just one curious click can be lethal to your computer.

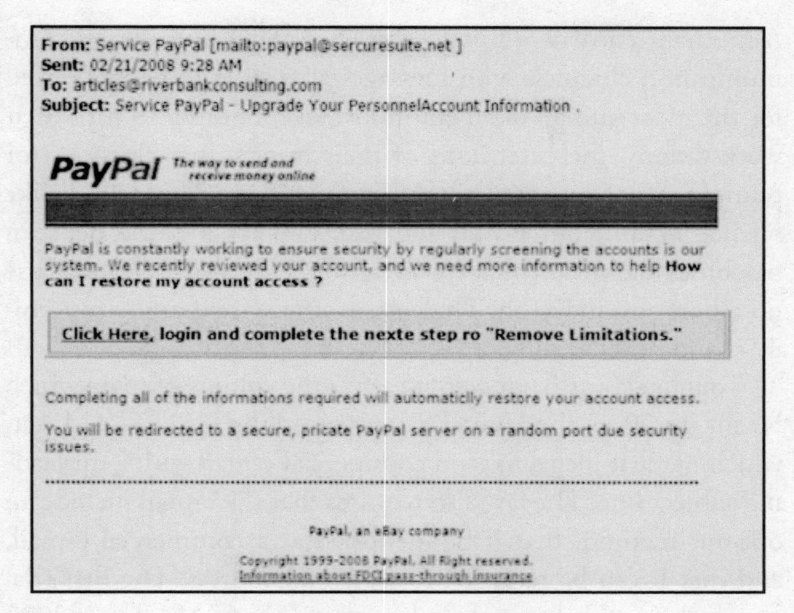

From: Service PayPal [mailto:paypal@sercuresuite.net]
Sent: 02/21/2008 9:28 AM
To: articles@riverbankconsulting.com
Subject: Service PayPal - Upgrade Your PersonnelAccount Information .

PayPal The way to send and receive money online

PayPal is constantly working to ensure security by regularly screening the accounts is our system. We recently reviewed your account, and we need more information to help **How can I restore my account access ?**

Click Here, login and complete the nexte step ro "Remove Limitations."

Completing all of the informations required will automaticlly restore your account access.

You will be redirected to a secure, pricate PayPal server on a random port due security issues.

PayPal, an eBay company

Copyright 1999-2008 PayPal. All Right reserved.
Information about FDCI pass-through insurance

A phishing e-mail scheme

Even worse, spammers have become very skilled at targeting people where they're most vulnerable. People love and need to be loved. Not surprisingly, the most successful phishing schemes mimic social networking sites, logging, in some cases, a 70 percent effectiveness rate. In May 2008, many of the most common malware spam e-mails—messages that provide a link to a Web site that triggers the download of software that will compromise a PC—came with the following headers: "Love You"; "With You By My Side"; "A Kiss So Gentle"; "Me & You." And then there's the bizarre: according to AOL, which blocked 1.5 billion spam messages a day in 2006, the most common junk e-mail subject line was "Donald Trump wants you, please respond."

Let's stop for a moment to ponder this curious condition.

A mechanized network of zombie PCs gangs up to flood communication channels with messages appealing to people's need for the most human of all emotions, love, in order to turn their workstations, the extensions of their minds, into factories for pumping out unwanted advertising messages. This sounds like science fiction, but it's far too real, and it's a battle between machines and people that we are losing. Like a virus, botnets use up a host and move on. The costs to the system are astronomical—and never ending.

Tougher laws do not seem to affect the volume of spam which is sent. In 2003, the United States passed the CAN-SPAM Act, which made it illegal to send commercial e-mail with a misleading subject line. The law also requires that the e-mail include an opt-out method, that it be identified as a commercial e-mail, and that it can be returned to an actual address. The first trial centered around charges based on the CAN-SPAM Act of 2003 didn't occur until 2007, when two men who had run a $2 million pornographic spam service were brought to trial for charges ranging from wire fraud to interstate traffic of hard-core pornographic images. The men used a server in Amsterdam to make it appear that the messages were coming from outside the country and registered their domain under the name of a fictitious employee at a shell corporation they had set up in the Republic of Mauritius. Each time someone clicked through a link in their e-mail spam to a pornographic Web site, the men received a commission. They also received a lot of complaints—over 650,000 were logged with AOL alone. In October 2007 they both received five-year prison sentences.

Their fine was a pittance compared to the $873 million judgment a New York District Court judge handed down against Adam Guerbuez, a Canadian man who used a phishing scheme to steal Facebook members' passwords. He then sent 4 million spam messages to Facebook members in the form of e-mails and postings on their walls, signed with the names of their friends, so they appeared to be legitimate. Among the products friends appeared to be endorsing were marijuana and the ubiquitous penile enhancement pills.

Solutions to the spam problem tend to have a short-lived life cycle, since spamming remains incredibly lucrative. A UC San Diego survey discovered that, with large enough networks of botnets behind them, spammers can become millionaires on a response rate of just 1 in 12.5 million e-mails. That hasn't stopped people from trying to prevent them from cashing in. In 2004, Bill Gates announced to the World Economic Forum, "Two years from now, spam will be solved." Gates turned out to be wrong, but you can hardly blame him for trying. In late 2004, it was reported that the most spammed person in the world was . . . Bill Gates, who received more than 4 million messages per year, most of them spam. "Literally, there's a whole department almost that takes care of it," said Microsoft CEO Steve Ballmer. They'll continue to be busy. Despite dips in spam, it's only going to increase. Studies have shown that online marketing is going to double from its current state by 2012. And the number of viruses in e-mail has been growing faster than the volume of e-mail itself.

Big Brother Is Watching

With all these threats to privacy, the message is clear: "You have zero privacy," Scott McNealy, the CEO of Sun Microsystems, once said about life on the Internet. "Get over it." Web sites you visit send tracking "cookies" to your browser, tiny parcels of text that go back and forth between a server and a client like your browser, allowing the Web site to store information about your preferences and the Web sites you have visited. Retailers mine your computer any time you purchase something online—and then turn around and sell it to the highest bidder. And it doesn't stop there. E-mail, which, it's important to remember, is stored on servers most of us don't own, is constantly monitored. Especially at work.

In 2001, 14 million U.S. workers—35 percent of the online workforce—had their Internet or e-mail under constant surveillance. Worldwide, 27 million workers were in the same boat. Employers spend $140 million a year on employee-monitoring software. Thanks to the Sarbanes-Oxley Act of 2002 and other regulations, publicly traded companies are required to archive their e-mail. Europe still has strong privacy protections for its employees, but many U.S. employers in the private sector, as long as they have an established policy and have put it into writing, can keep a close eye on what their employees send and receive, and where they point their browsers. "Some companies say they do it to control the information that employees send through the corporate network," wrote Matt Villano in *The New York Times*. "Other companies do it to make sure employees stay on task, or as a measure of network security. Other companies monitor e-mail to see how employees are communicating with customers." Mary Crane, the president of a consulting firm in Denver, gave this advice: "The last thing you want to do is make your employer think you're slacking off. . . . Nothing you're doing on e-mail is worth jeopardizing your career." The names of the companies that specialize in employee monitoring will make you never want to goof at work again: ICaughtYou.com, eSniff, Cyclope Series.

It's not just your employer peeping in, however. Lovers and spouses do it, too. A survey done in Oxford revealed that one in five people had spied on their partner's e-mails or texts. Cheaters are constantly caught. "Spurned lovers steal each other's BlackBerrys," wrote Brad Stone. "Suspicious spouses hack into each other's e-mail accounts. They load surveillance software onto the family PC, sometimes discovering shocking infidelities." In one case, Stone described a woman who was convinced her husband was straying—he was far too obsessed with his BlackBerry. On his birthday she drew him a bubble bath and rifled through his handheld while he was soaking, discovering that he did have a bit on the side and planned to meet her *that night*. All this evidence gleaned from glowing devices winds up in divorce proceedings, where the electronic paper trail becomes the knife you stick in your former partner's back. "I do not like to put things on e-mail," said one divorce lawyer. "There's no way it's private. Nothing is fully protected once you hit the send button."

Increasingly, governments are getting into the act. And it's hard not to understand why. "In this new kind of war," James Bamford has written about the importance of Internet communications for stateless agents, "in which motels are used as barracks and commercial jets become powerful weapons, public libraries and Internet cafes are quickly transformed into communication centers." But surveillance of e-mail predates the so-called war on terror. Indeed, long before it was revealed that Mohamed Atta and the other 9/11 hijackers had communicated by e-mail, logging in at public libraries, no less, the U.S. National Security Agency (NSA) was keeping tabs on "chatter" by monitoring e-mails. The scope of its listening capacities is truly awesome. As James Bamford describes in *A Pretext for War,* "dozens of listening posts around the world each sweep in as many as two million phone calls, faxes, e-mail messages, and other types of communications *per hour*."

In December 2005, though, *New York Times* reporter James Risen and Eric Lichtblau broke the story that President Bush had told the NSA, whose mandate is to spy on foreign and foreign agent communications, to spy without warrants on the phone calls and e-mails of people *inside* the United States. President Bush assured the American people that one end of the conversation had to be outside the United States, but in fact the NSA was "trolling through vast troves of Americans' telephone and e-mail conversations with artificial intelligence," writes Maureen Webb, a human rights lawyer and activist, "looking for key words and patterns."

Most alarming for many Americans is the fact that the communications companies were helping them. To a certain degree, this is not new. According to a federal statute called the Communications Assistance for Law Enforcement Act (CALEA), passed in 1994, communications companies must design their facilities so that their network can be easily monitored. As Bamford explains in *The Shadow Factory,* "it even requires the company to install the eavesdropping devices themselves if necessary and then never reveal their existence."

As Webb describes in her book *Illusions of Security,* this is just the tip of the iceberg in a vast data-mining program that will draw from multiple databases to create a "black box," a giant trove of data vastly more sophisticated than the black boxes that Internet service providers have to monitor user traffic.

> Little is publicly known about the workings of these, except that they are like the "packet sniffers" typically employed by computer network operators for security and maintenance programs running in a computer that is hooked into the network at a location where it can monitor traffic flowing in and out of systems.
>
> The Internet, which used to be a law enforcement nightmare, has become a useful tool now that

> technology has developed to monitor it. Sniffers can monitor the entire data stream, searching for keywords, phrases, or strings such as net addresses or e-mail accounts. They can then record or retransmit for further review anything that fits their search criteria. Black boxes are apparently connected directly to government agencies by high-speed links in some countries.

In other words, they're vacuuming up e-mails at broadband speed as you read this. In 2002, the NSA was gobbling up 650 million intercepts a day, the same number of pieces of mail the U.S. Postal Service then delivered in America every day.

On the Perils of Driving Fast

> *We believe at this point, we need people to change their behavior. . . . We need to make sure when people are getting on the highway, they are prepared to travel safely.*
>
> —John Njord, director of the Utah Department of Transportation

One of the deadliest places you can be in the United States is on a roadway. In 2006, 42,642 people were killed in traffic accidents; it was a banner year since it represented a drop of roughly 2 percent from 2005. Yet the Interstate Highway System and the smaller roadways have become essential to living in the United States. People get in their car, buckle up or don't, and drive off without a flicker of a thought that they might be on their way to becoming one of the 116 fatalities about to happen that day. The urban planning that developed as a result of roadway use often doesn't give them much choice; they have to drive.

It didn't have to be this way. The U.S. Interstate Highway System, the largest public works project in the history of the world, was born out of the same military background as ARPANET. President Dwight Eisenhower signed it into law on June 29, 1956, having been heavily influenced by traveling around Germany on the Autobahn after World War I and realizing that America didn't have a similar system for mobilizing and transporting troops. It also happened to help people travel around the country and was a great boon to automakers, who lobbied heartily for it.

For every benefit of the interstates—and there have been many, from faster shipment of goods to greater ease of travel and an enormous number of jobs for Americans—there have been many side effects. Interstates have torn up the landscape, destroyed the frontier, made Americans dependent on automobiles that pump tons of carbon emissions into the atmosphere, and created a system of suburban living that siphoned tax wealth out of cities, causing slums to develop and flourish—one of the most potent examples of which is the ring around Los Angeles. "Nothing has been learned from the dismal California experience," Mike Davis once wrote, "not even the elementary lesson that freeways increase sprawl and consequently the demand for additional freeways."

In 2009, when the Pennsylvania Turnpike/Interstate 95 Interchange Project is finished, the last project of the original interstate plan will be completed, but we are nowhere near the end of the infinite highway system coconceived in Eisenhower's era by the good folks at ARPA and now known as the Internet. Each month, new traffic circles, abutments, and auxiliary lanes are thrown up and quickly exploited as new host computers go online, new devices are invented, more and more people use wireless. The speed limit gets higher and higher.

Entire virtual cities are thrown up and torn down, and the speed of our ability to travel between them virtually means that we actually don't have to physically leave the house anymore. For this reason, a Stanford professor, Norman Nie, has commented, "The internet could be the ultimate isolating technology that further reduces our participation in communities even more than did automobiles and television before it."

And what of this highway? There is so much to see by the roadside, so much of enormous use, be it the endless library of Google books, the streaming sports scores on ESPN, or the river of talk bisecting the screen after an episode of *America's Next Top Model.* But as we just learned, if we think of our e-mail accounts as the middle lane on this interstate, what kind of road is this? Would anybody in real life get onto a freeway where nine out of ten cars had a penile enlargement ad on the side, let alone the occasional thief looking to siphon off your financial information or breach your home, where the identity of drivers could be concealed, where the government or your employer or your lover was constantly monitoring you, where people veered into your lane to impose their needs on you two hundred times a day? What sort of ballistic defensive driving course could prepare us for this kind of travel?

Movement is metaphor, because all travel—virtual or actual—changes us. The Greek word for "carry" was *metaphora*. What we bring and carry into the virtual world, and what we put up with there, changes us. It alters our priorities, shrinks or expands our empathic bandwidth. We know this about e-mail; keeping up with it, as useful as the message system is in so many other ways, just like that highway, is shrinking our focus. It's one of the reasons why it causes such anxiety. None of us can change the system on our own. And then there's a more insidious reason: we simply cannot stop ourselves.

4

THIS IS YOUR BRAIN ON E-MAIL

If the medium is the message, what does that say about new survey results that found nearly 60 percent of respondents check their e-mail when they're answering the call of nature?

—Michelle Masterson, Channel Web

Now that handheld devices give us 24/7, virtually worldwide access to e-mail, there is nowhere, it would seem, that people do not pause to check it. We log on during the drive to work, download a few messages on the train ride home; we look at it in the bath and in between sermons at church. Sixty-two percent of Americans check their e-mail on vacation and respond to work queries, at a time when they're supposed to be relaxing. According to *Merriam-Webster's Collegiate Dictionary*, "vacation" means "a respite or a time of respite from something" or "a scheduled period during which activity (as of a court or school) is suspended." Nothing is suspended in the wired vacation of the twenty-first century. Any time there's a moment of silence, a break between moments, e-mail insinuates itself with stunning regularity. "You know those pregnant pauses you have on elevators? That's a great time to pull out a BlackBerry and get some work done," says Raul Fernandez, the CEO of Dimension Data North America.

There is no downtime anymore, even at bedtime. Sixty-seven percent of the four thousand people age thirteen and over surveyed in AOL's 2008 e-mail addiction poll admitted to having checked e-mail in their bed, in their pajamas. In the 1996 film *She's the One,* Jennifer Aniston is married to a distracted financier who cares more about his job than his wife; we know this because he takes his laptop to bed. Now many of us are doing the same, even if our devices have shrunk along with our trust in financiers. Sean Young of Phoenix is one. He logs on before and after the gym, by the pool, in the car, and leaves his handheld inches from his face at night so he never misses a message. "I just realized I have a problem," Young said, describing his daily routine of message consumption in an e-mail to a reporter.

He's not alone. *Nearly half the people* in AOL's survey claimed that they were addicted to e-mail. The technology that was supposed to set us free to work from anywhere, to check in and clock out on our own time, has now become the longest employee leash ever invented because we can't seem to log off. We haven't just tried to merge with the machine, to marry the damn thing; it has become our iron lung. "I have friends and relatives that carry BlackBerrys with them 24 hours a day, fully prepared to drop anything in their lives and work at a moment's notice," wrote Tim O'Leary, the CEO of a marketing firm. "I'm tethered to my laptop as if it were an oxygen machine I must cart around to keep me breathing." The word "crackberry" was *Webster's New World College Dictionary*'s 2006 word of the year.

The most "addicted" metropolis in America is, not surprisingly, New York, the city that never sleeps—and apparently never stops clicking: 50 percent of Gothamites feel they are addicted to their e-mail. Lunch hour in Manhattan can sometimes feel like an outtake from a strange daylight zombie film: e-mail drones, flicking and scrolling through their handhelds, checking e-mails that they could just as easily read twenty minutes later at their desk, are given a wide berth on city streets by the not-yet-addicted.

There are several reasons for this burgeoning obsession. Mail has always traveled to us with a small but palpable comet trail of anticipation. Regular delivery of the post created a daily rhythm of expectation. We know that bills and official forms will come. But there might also be postcards from friends, Christmas cards, magazines, or maybe more. In 1967, the direct marketing firm Publishers Clearing House launched a prize giveaway. It might not just be your subscription of *Runner's World* in your mailbox; it might also be a $1 million check with the "prize patrol" in tow.

Now that our inboxes have become both our most used mailbox and virtual doorstep, it's hard not to have the same complicated mixture of good and bad expectations when checking e-mail. Except that we no longer have to wait. The BlackBerry was introduced in 1999 and by 2004 had 1 million users, a number that doubled ten months later. As of June 2009, that number had reached 28.5 million worldwide, and that doesn't even count people using e-mail-enabled cellular phones. Millions upon millions of people the world over now can and do constantly access their e-mail. Psychologists have discovered that their behavior in doing so is very like that of people sitting before a slot machine.

Neurologists now understand why these standbys of casinos are addictive: they work on a principle called variable interval reinforcement schedule, which Tom Stafford, a lecturer in the Department of Psychology at the University of Sheffield, explained has been established as the way to train the strongest habits. "This means that rather than reward an action every time it is performed, you reward it sometimes, but not in a predictable way. So with email, usually when I check it there is nothing interesting, but every so often there's something wonderful—an invite out, or maybe some juicy gossip—and I get a reward."

There are chemical reasons for why this reward feels good, reasons that go beyond the quality or rarity of the gossip. The midbrain is constantly trying to make predictions about when we will and won't be rewarded. Brain imaging is beginning to show that when we get a big reward—such as a jackpot payout—dopamine, the hormone and neurotransmitter, floods the anterior cingulate, the part of the brain that appears to control mechanical functions such as heartbeat and breathing, as well as rational functions such as decision making and reward anticipation. If we're performing an action that doesn't always pay out, but does *some of the time,* such as playing the slots, the lesson learned is that if we want a reward we need to keep pulling that lever.

So it is with our e-mail. We need to keep clicking that send/receive tab—even when our computer is set to automatically check e-mail every ninety seconds—to get the reward we've come to expect will arrive sooner or later. *Someone is thinking of me*. The addictive nature of working in this environment has been good for response rates. In one recent survey, it took people an average of just one minute and forty-four seconds to respond to an e-mail pop-up alert on their computer. Seventy percent of alerts provoked a reaction in just *seven seconds*.

As with any vice, it's a disaster when you take e-mail away, even if for only a few hours. When the BlackBerry network went down for several hours one night and into the next day in 2007, it deprived 8 million users of their wireless e-mail. Many of them panicked. "My blood ran cold," said one real estate consultant, who was traveling on business. "I was offline." In the summer of 2008, Google's popular Gmail service went down for just a few hours, and the company was flooded with responses. "We feel your pain and we're sorry," the company wrote on its blog. Going offline causes huge amounts of stress for companies, especially small ones. A survey done in England revealed that "77 percent of office workers and company owners agree that e-mail downtime causes major stress at work." Forty percent responded with agitated mouse clicking. Ten percent physically assaulted their computers. Postcards may have been a craze, but there's nothing that even compares to this level of devotion to e-mail.

The physiological qualities of e-mail dependency, if they don't grow out of a psychological dimension, can soon acquire one, as with any chemical dependency. "If I didn't hear 'beep-beep' every time I turn on the computer," said one senior citizen who adopted e-mail in 1994, "I'd die." E-mail has become a way to be reminded that we exist in a world overloaded with connections, that we are needed. Out of the Internet we have constructed a new communications environment that enables us to constantly feed that need—to be plugged in, surrounded by links to all of our friends and colleagues. Artemio Ramirez, Jr., an assistant professor at the Hugh Downs School of Human Communication at Arizona State University, points out that e-mail addicts are people who like to feel desired, and needed, which, as the statistics bear out, is a lot of us. "It makes us feel as part of a community or network," Ramirez said. It's a basic human desire—yet the way that e-mail has speeded it up has destroyed our ability to want much else.

For these reasons, some psychologists are pushing to have "Internet addiction" be broadly classified as a clinical disorder. Dr. Jerald Block of the Oregon Health and Science University is one of them, and he says that sufferers show all the classic signs of addiction. They forget to eat and sleep, require more advanced technology and higher doses—in this case, a larger volume of e-mail, a constant connection to it—to get their fix. But they are in for perpetual disappointment. "When we log in to our e-mail server," writes Richard DeGrandpre in *Digitopia,* "the expectation of finding new mail negates any possible excitement or surprise; if there's no new mail, we're disappointed." So we check it more and more. As the condition progresses, sufferers feel increasingly isolated from society, become argumentative, and fall into depression. They spend time gaming online, looking at news and pornography, and e-mailing. Early sufferers, Block says, tended to be highly educated, socially awkward men, but now more and more they are middle-aged women who are either at home alone or working. In fact, there's no better place for an Internet addict these days than at work.

Working in a Climate of Interruption

> *In the era of e-mail, instant messaging, Googling, e-commerce and iTunes, potential distractions while seated at a computer are not only ever-present but very enticing. Distracting oneself used to consist of sharpening a half-dozen pencils or lighting a cigarette. Today, there is a universe of diversions to buy, hear, watch and forward, which makes focusing on a task all the more challenging.*
>
> —Katie Hafner

> *What information consumes is rather obvious: it consumes the attention of its recipients. Hence a wealth of information creates*

a poverty of attention, and a need to allocate that attention efficiently among the overabundance of information sources that might consume it.

—Herb Simon

We work in the most distraction-prone workplace in the history of mankind. We can be reached on the phone, by fax, instant message, Facebook, text message, cellular phone, letter, and occasionally in person. Throughout the day, for many people and especially for the very busy, these various channels and machines are blinking and beeping like an ambulance trying to cross a busy intersection at rush hour. In 2006, one study found that the average U.S. office worker was interrupted eleven times an hour. The cost of these interruptions, in which e-mail plays a large role, runs close to $600 billion in the United States alone.

We live in a culture in which doing everything all at once is admired and encouraged—have our spreadsheet open while we check e-mail, chin the phone into our shoulder, and accept notes from a passing office messenger. Our desk is Grand Central and we are the conductor, and it feels good. Why? If we're this busy, clearly we're needed; we have a purpose. We are essential. The Internet and e-mail have certainly created a "desire to be in the know, to not be left out, that ends up taking up a lot of our time"—at the expense of getting things done, said Mark Ellwood, the president of Pace Productivity, which studies how employees spend their time.

The evidence is in, though, and we can't multitask the way the technology we use at work leads us to believe we can. Our brains are what's telling us this. "Multitasking messes with the brain in several ways," Walter Kirn wrote in an essay called "The Autumn of the Multitaskers" in 2007. "At the most basic level, the mental balancing acts that it requires—the constant switching and pivoting—energize regions of the brain that specialize in visual processing and physical coordination and simultaneously appear to shortchange some of the higher areas related to memory and learning."

A UCLA experiment bore this out. A group of twentysomethings were asked to sort a deck of cards—once in silence, a second time while listening to randomly selected sounds in search of specific tones. "The subjects' brains coped with the additional task by shifting responsibility from the hippocampus—which stores and recalls information—to the striatum," Kirn explained, "which takes care of rote, repetitive activities. Thanks to this switch, the subjects managed to sort the cards just as well with the musical distraction—but they had a much harder time remembering what, exactly, they'd been sorting once the experiment was over."

In other words, a work climate that revolves around multitasking and constant interruptions has narrowed our cognitive window down to a core, basic facility: rote, mechanical tasks. We like to think we are in control of our environment, that we act upon it and shape it to our needs. It works both ways, though; changes we make to the world can have unseen ramifications that impact our ability to live in it. And research is revealing that our use of technology has begun to alter our attention span; we've started reverse engineering our brains for speed, as opposed to mindfulness. It is perhaps for this reason—aside from its continuing cultural marginalization as an irrelevant art form—that poetry, despite being short and bite-sized, has not become a national pastime in the age of the Internet. As Donald Revell has pointed out, poetry is not about brevity or length but about attention, and "attention is a question of entirety, of being wholly present."

The longer we work this way, the harder it's going to be to do things that force us out of our reactive, dronelike existence, such as reading a novel or even a long magazine piece. "Immersing myself in a book or a lengthy article used to be easy," wrote Nicholas Carr in an essay entitled "Is Google Making Us Stupid?" "My mind would get caught up in the narrative or the turns of the argument, and I'd spend hours strolling through long stretches of prose. That's rarely the case anymore. Now my concentration often starts to drift after two or three pages. I get fidgety, lose the thread, begin looking for something else to do."

There's a reason for this: Reading and other meditative tasks are best performed in what psychologist Mihaly Csikszentmihalyi called a "state of flow," in which our focus narrows, the world seems to drop away, and we become less conscious of ourselves and more deeply immersed in ideas and language and complex thought. Many communication tools, however, actually inhibit this state. "Telephones and tape recorders, computers and fax machines are more efficient in conveying news," Csikszentmihalyi wrote in 1991, before e-mail had become as pervasive as all of these tools. "If only the point to writing were to *transmit* information, then it would deserve to become obsolete. But the point of writing is to *create* information, not to pass it along."

A whole new field of attention studies has emerged to capture the shift that has already occurred in how the brain works. One of the most intriguing recent findings in this area came from a Swedish cognitive scientist, Torkel Klingberg, who posited in *The Overflowing Brain* that there are roughly two types of attention: one a controlled, task-oriented kind, such as that required to crunch a spreadsheet; the other a stimulus-driven kind of attention, such as the way a car horn causes us to whip our heads around and prevents us, in most cases, from stepping out into a busy street. Klingberg connects these two types of attention to two kinds of memory: short-term working memory and long-term memory.

The implications of this model on modern working life are enormous, as the greatest challenge presented by working in an environment of constant stimuli is maintaining focus, as well as keeping a lifeline to our working memory. One two-year study with children Klingberg cited showed that working at overload capacity does, in fact, improve working memory and possibly even problem-solving skills. Multitasking may not be perfect, but it can push the brain to add new capacity; the problem, however, remains that the small gains in capacity are continuously, rapidly outstripped by the speeding up and growing volume of incoming demands on our attention. The center, it turns out, just won't hold; so we create and push out into this world a kind of orienting capsule: ourselves.

The Digital Self: Introducing a Whole New You

No modern technology since the automobile has been more associated with our selves—the freedom to be, to explore, to augment and design ourselves—than the Internet. If you bought a plum-colored Dodge Charger in 1970, you were expressing, in a not very subtle way, that you were no shrinking violet. We now have a similar, much cheaper opportunity for transformation just by connecting to the Internet. Which "profile" do you want to use on your home PC? Which e-mail account? Do you want to be *AmyWilliams1972@yahoo.com*? Or would you rather be *surferchick1972@yahoo.com*? Depending on what circles this (fictional) Amy Williams traveled in, perhaps it would be better to go as Amy.Williams1972@fas.harvard.edu—as long as she was actually, of course, an alumna of Harvard University.

Our addresses, in a simple way, say who we are and what we do, what we care about. They are a starting point of an interaction even more than a name since they can be chosen and manipulated. One study of 599 accounts in Germany showed that judgments made about people based solely on their e-mail addresses were ballpark correct and that observers made subtle distinctions from one address to another.

These addresses are just a very small signal—like the turning of leaves before a coming storm—of the enormous change brought about by the Internet with regard to our notions of a self. The rise of Internet-based communications, communities, and entire identities has become a fascinating way to watch people behave and negotiate their new, unhinged digital selves. The medium, as it turns out, isn't just the message; it's a hall of mirrors. In psychoanalytic terms, writes John Suler, a professor of psychology at Rider University, "computers and cyberspace may become a type of 'transitional space' that is an extension of the individual's intrapsychic world. It may be experienced as an intermediate zone between self and other that is part self and part other. As they read on their screen the e-mail, newsgroup, or chat message written by an internet comrade, some people feel as if their mind is merged or blended with that of the other."

What Rider is describing is called, in psychological terms, merger transference, and it starts, as I mentioned in the previous chapter, with the computer itself. Our computers have become extensions of ourselves—of our inner space—housing our personal letters and work and music and movies and most private financial information, sometimes all in one place. The visual cues that we have left our "desktop" and are now "out" on the information highway when we launched a browser, though, are quite small; we still feel "in" there, inside our extended head. So it's extremely easy to feel as if somehow everything we do and everyone we interact with will play out inside us as well.

For those who struggle with this boundary, the computer and working via e-mail have made us, in a sense, narcissists. In his essay "On Narcissism," Sigmund Freud proposed two ideas of the narcissist: one revolved around the concept of self-love, the other stemmed from a state of mind that has no awareness that the self and other exist. Working over the computer and via the Internet, we have numerous cues that feed both instincts. Many of the most popular Web sites cater to our idea of selfhood and agency: MySpace.com; YouTube; YouPorn.com (don't laugh, it's routinely in the top thirty most visited sites in America).

Working over e-mail also reinforces this sense of being at the center of things—with a complicated caveat. The nature of the computing interface and the emotional attachment we forge to the machine make it harder and harder to remember that this voice, this text on the machine has come from outside of us. Suler explains that for some Internet users, "reading another person's message might be experienced as a voice within one's head, as if that person magically has been inserted or 'introjected' into one's psyche."

Thus, criticism that zings in over e-mail can feel like shouting; it can also flush out our deepest self-doubts and begin to sound more like an echo of ourselves than a comment from outside. We have several defenses against this—we can fire right back or flame, which I will discuss later. Or we can dissemble, improvise, or blend into the background. It used to be that clothes made the man, but for the hours we're on the Internet or typing to one another, that's not so. We can be whoever we want to be on a scale that's never before existed or been quite so easy.

JOHN FREEMAN

Writing Our Way into Existence

> *Email allows me to indulge my new meditative technique: annihilation via impersonation. I answer each letter in my interlocutor's voice, and forty responses later I am no one and everyone.*
>
> —Don Paterson, *The Blind Eye*

> *In cyberspace I can change myself as easily as I change my clothes. Identity becomes infinitely plastic in a play of images that knows no end. Consistency is no longer a virtue but becomes a vice; integration is limitation. With everything always shifting, every one is no one.*
>
> —Mark Taylor and Esa Saarinen, "Shifting Subjects 1"

What does it mean that you can have as many addresses on the Internet as you like, for free? That they will become an integral part of your daily life? One could argue that this isn't an entirely new development. Anonymous letters were an issue in the past, after all, and early writers frequently wrote under a pseudonym, as do some modern writers. Benjamin Franklin made his literary debut by slipping articles under his newspaper-publishing brother's door in the name of Silence Dogood; later on in life he published *Poor Richard's Almanack* under the name Richard Saunders. C. S. Lewis published poems under the name Clive Hamilton to protect his reputation as an Oxford don. The seventeenth-century Japanese poet Matsuo Kinsaku used fifteen different *haiga,* or pen names, before he settled simply on Basho, which was the name of a banana plant. Mary Ann Evans used a male pen name because she wanted her work to be treated seriously: we know her as George Eliot.

It's not an accident that all of these examples come from writers. Until the end of the twentieth century, they were the only people with the ability to put their thoughts into writing and have them distributed to a large and ready audience. One could write a letter to the newspaper and hope the editors printed it; bang away at a novel and send it off to a publisher and wait by the mailbox for a reply. But the vast majority of people wrote to communicate person to person, via letters and postcards and occasionally by telegram.

Now we can write for the world, or we can write to a friend. If both are posted on the Web in the form of a blog, perhaps the easiest way to publish ever created, it will be up to netizens' search habits as to whether either is read. Similarly, the forwardability of e-mail means that these intentions are easily blurred: an e-mail to a friend may, if clever or embarrassing enough, be read by hundreds of thousands of people. An e-mail to a large group may not be read by any of them.

The range of this possibility and the volume of text we're creating mean that more and more people are experiencing the metaphysical and stylistic dilemma once peculiar to writers. We write our way into being. Having different identities, different voices, different e-mail addresses—which is recommended if one wants to outrun spammers—gives us a degree of agency and control over the various online environments we visit. Disguised or obscured, we can fabricate; as our professional selves, we can do business. Using an e-mail address specifically to purchase items protects us from having our personal or work e-mail inboxes from being inundated with spam. As the Russian theorist Mikhail Bakhtin would point out, though, all these identities are unstable. The "I" that means "you" is a social construction just as much as surferchick1972.

The Internet has placed this fluidity in the most receptive

medium yet. Identity roulette has been part of the Web since its early days, when people logged on to Multi-User Dungeons (MUDs) in the 1970s to play an online textual version of Dungeons and Dragons. Chat rooms allowed people to be whoever they wanted to be—even though one of the first things asked of participants was "Morf" (male or female?) or, in other situations, "Sorg" (straight or gay?).

E-mail, as it became the primary screen-to-screen, person-to-person form of discourse, has domesticated this textual identity game. A study carried out on schoolchildren in Yorkshire, England, revealed that children learn this—and the need to represent it—immediately and take advantage of it. One of them took to using the signature "xxxxxxx—*Kavita*—xxxxxxxx." Asked later why, she responded, "I do that all the time . . . when I do my name on the computer and stuff. . . . I put my name like that cos it looks really neat and it looks boring just as a name."

We construct these identities visually and textually over e-mail because we can—but also because we must. As the volume of communication via e-mail increases, we need to differentiate ourselves, make our voices heard, cut through the noise. The prerogatives and seductions that once were the concern of writers, who had to keep us turning the pages, are now ours as correspondents. A distinctive voice is invaluable among friends and in business, just as it was in telling a story. "When you read a novel the voice is telling you a story; when you read a poem it's usually talking about what its owner is feeling; but neither the medium nor the message is the point," A. L. Alvarez writes in *The Writer's Voice.* "The point is that the voice is unlike any voice you have ever heard and it is speaking directly to you, communing with you in private, right in your ear, and in its own distinctive way." It's not just a matter of voice, though; research has shown that people imagine their lives in stories and tend to live out their days in keeping with the narratives they tell. "The way people replay and recast memories," wrote Benedict Carey, "day by day, deepens and reshapes their larger life story."

The newly democratized narrative and authorial power granted by e-mail is an unprecedented freedom and responsibility compared to previous eras of communication. The Internet has created a situation where you can be many things at once and live behind as many false identities as you choose; creating such identities is even recommended to protect your privacy. "Never give your true identity when signing up for something online," Fred Davis, the founder and CEO of Lumeria, an Internet privacy firm, told a *San Francisco Chronicle* reporter. The writer later added, "Another way to mask your identity is to set up free e-mail accounts, like those available at a portal site (such as Yahoo or Excite) or at Hotmail, and use them in identifying yourself around the Web."

The practice is widespread. In 2006, a writer at the *Los Angeles Times* was suspended after it was discovered that he had been posting comments under fake names, Mikekoshi and Nofanofcablecos, on his blog and those of others. "Can a company that derives economic value from its reputation for literacy, judiciousness and taste comfortably lend its imprimatur to an unfiltered online diary?" the reporter wrote in his column and on the blog, skeptical from the beginning of the experiment in the new online form. "Blogs are by nature almost impossible to censor."

Events like this—and the ease with which one can fabricate new online identities for e-mail or blogging—makes trust over the Internet difficult to develop and easy to smash. This destructive cycle starts early, in school, where kids have always been picked on face-to-face. Now children have to suffer in the virtual world, too. Sixty percent of cyberbullying is anonymous, and many times it comes from friends who know passwords and secrets. One of the most devastating examples occurred in Missouri. A thirteen-year-old named Megan Meier was contacted by a boy named Josh Evans online through the popular social networking site MySpace.com. "Mom! Mom! Look at him!" Meier said to her mother, pointing at his profile, which included the picture of a young, bare-chested boy with stylishly rumpled hair, asking permission to add him as a friend. Meier's mother said yes, and soon a friendship developed over MySpace messages, which are like Web-based e-mails. Meier had been depressed, and her spirits lifted. Here is how she described herself:

M is for Modern
E is for Enthusiastic
G is for Goofy
A is for Alluring
N is for Neglected.

Then, just as suddenly as Meier's neglect had ended, the messages from her new friend changed. Josh wrote, "I don't know if I want to be friends with you anymore because I've heard that you are not very nice to your friends." Other messages were blunter. Then postings appeared on bulletin boards: "Megan Meier is fat"; "Megan Meier is a slut." The final message was downright cruel: "Everybody in O'Fallon knows how you are. You are a bad person and everybody hates you. Have a shitty rest of your life. The world would be a better place without you."

On October 16, 2006, in a tailspin after this message, Meier hanged herself. She died the next day. Six weeks later, it turned out that "Josh Evans" was actually a forty-seven-year-old woman named Lori Drew, the mother of a friend with whom Meier had lost touch after changing schools. Drew had apparently concocted the identity as a hoax and even laughed about it with other neighbors, enjoying the ability to "mess with Megan." In the fall of 2008, Drew was convicted of just three misdemeanor charges of accessing computers without authorization. The jury declined to convict her on a felony charge of accessing a computer without authorization to inflict emotional distress.

What Meier experienced, in an intimate, destructive, and deeply personal way, is an aggressive form of flaming—hostile or insulting communication over the Internet. It even continued after her death; a blog emerged after her death with postings claiming she'd had it coming. The anonymity afforded by the Internet has made it all too easy to criticize or roast someone publicly or right in their inbox. Like identity gymnastics, it's been with the Internet from the beginning. "Flame messages often use more brute force than is strictly necessary," writes Virginia Shea in *Netiquette,* "but that's half the fun." In other words, it's a game made for "teaching lessons" that often gets out of hand.

Trying to parse digs from duds, frenemies from simply frantic e-mailers, is a very difficult task in an environment so permeated with anonymity, speed, and lack of face-to-face interaction. Thinking too much about it can lead to a constant swirl of paranoia. Is it *me,* one wonders, or was that last message rude? The blogger "Elkins"—a pseudonym, by the way—astutely observed the mental self-questioning that takes place due to aggression among women on fan boards and the diabolical but very real possibility that this kind of abuse is meant to inspire such confusion, like Lori Drew "messing" with Megan Meier. Elkins's comment could apply to aggression in any form of computer-mediated conversation, however, including e-mail:

> There's definitely a "Gaslighting" effect to aggression which is so often denied: it serves to make the target doubt her own perception of reality. If it *seems* as if someone is trying to hurt you, but when confronted the person in question denies that this was at all the intent, then how do you respond? Whom do you trust? After all, you *could* have misinterpreted, or overreacted; and since it's quite often a purported "friend" aggressing against you in this fashion, you really wouldn't want to level a false accusation. Yet it's hard for the target of, say, an extended whispering campaign to avoid the conclusion that people really are out to get her because . . . well, because actually? They *are*.

As more and more people get connected, though, the Internet has become riddled with these ecological/emotional fires. Every interaction, it seems, has the potential to become a flame war. People's inability to understand tone in e-mails can lead to it; so can a bad day. There's another factor, however, that explains why people can be so rude to one another online over e-mail. We're all working in a medium that encourages disinhibition.

> *Flaming can be induced in some people with alarming ease. Consider an experiment, reported in 2002 in* The Journal of Language and Social Psychology, *in which pairs of college students—strangers—were put in separate booths to get to know each other better by exchanging messages in a simulated online chat room.*
>
> *While coming and going into the lab, the students were well behaved. But the experimenter was stunned to see the messages*

> *many of the students sent. About 20 percent of the e-mail conversations immediately became outrageously lewd or simply rude.*
>
> —DANIEL GOLEMAN, *The New York Times,* 2007

Two financial consultants who don't know each other are corresponding over e-mail to set up a meeting. The woman signs off with her contact information, asking, "Is there anything else I can provide?" Her cohort, whom she has just "met," replies, "How about a picture?" A book editor receives an e-mail from someone he met once that begins "It's been a hard year for me" and ends with details of the writer's mental breakdown. A manager of artists e-mails a booker to say his singer cannot appear because she has fallen ill. Facing a hole in his schedule, the booker fires off an angry reply: "If she is sick, which I sincerely *doubt,* I wish her a speedy recovery, but let her know she has really put us out."

These are all examples of what psychologists call disinhibition—a filter drops, and we write things we probably wouldn't say to another in person, at least not after such a brief acquaintance. No environment induces it quite as easily as computer-mediated communication. Indeed, the PC may have extended the human mind, but it's missing a few key human circuits that modulate social interaction. Neurologists now know that many of the key mechanisms of communication reside in the prefrontal cortex of the human brain. "These circuits instantaneously monitor ourselves and the other person during a live interaction," wrote Daniel Goleman on www.edge.org, "and automatically guide our responses so they are appropriate and smooth." One of the key tasks of these circuits is inhibiting "impulses for actions that would be rude or simply inappropriate—or outright dangerous."

But awkward as it sometimes feels to be inside it, our body, as it turns out, is the best, most sophisticated interface for appropriate communication. It has multiple valences; it has smell, touch, taste, and sight. It allows us to keep ourselves in check by providing real-time, continuous feedback from another person: facial expressions, the slightest twitch of an eyebrow, gestures, pauses, eye contact, the squeeze of an arm. Our bodies often embarrass us. "Man is the sole animal whose nudity offends his own companions," wrote Montaigne, "and the only one who, in his natural actions, withdraws and hides himself from his own kind."

The desire to transcend our fleshly envelope, to find a purer, more seamless form of talking and being is understandable—Emerson's transparent eyeball was in fact an extension of that wish. Communication technology, however, has been catering to that desire with increasing ability, from the telegram to the telephone, even if it is, in an idealized way, putting us right back into the *idea* of our bodies. It's a temporary relief, as we're discovering, sometimes not one at all. In Nicholson Baker's novel *Vox,* a man and a woman talk over a sex hotline. "I called tonight I think out of the same impulse," she says, "the idea that five or six men would hear me come, as if my voice was this *thing,* this disembodied body, out there, and as they moaned they would be overlaying their moans onto it, and, in a way, coming onto it, and the idea appealed to me, but then when I actually made the call, the reality of it was that the men were so irritating, either passive, wanting me to entertain them, or full of what-are-your-measurements questions, and so I was silent for a while, and then I heard your voice and I liked it."

There's a paradox here to this woman's experience on the chat line. The disinhibiting factors of the telephone that allow her to perform her own orgasm before a group of strangers for the same reason also work on the other participants on the line:

they can bark requests, relax in passivity, measure and assess nakedly and publicly, but without having to be seen. The same is true for written interactions over the Internet, but even more so. As Goleman says, "The Internet has no means to allow such realtime feedback (other than rarely used two-way audio/video streams). That puts our inhibitory circuitry at a loss—there is no signal to monitor from the other person. This results in disinhibition: impulse unleashed."

Another explanation of disinhibition leads back to the brain, having less to do with our "filters" and more with nuts-and-bolts functionality. One of the fastest-growing areas of neuroscience is the study of mirror neurons, highly active cells in our nervous system that, when we watch an action performed by another person, "fire," sending pulselike waves of voltage across cell membranes and creating the sensation that what we are watching is actually something we are doing or experiencing ourselves. Transcranial magnetic stimulation—the sending of low-voltage electric charges to parts of the brain to study its functionality—has confirmed this research.

The study of mirror neurons is still developing, but it is beginning to shed light on motor and language development, and also empathy. We may cry at the sight of a sad friend, screw up our face when we see someone react to a bad smell, or cringe when we see someone punched, because we are mirroring what she is experiencing. Research by the French-German neuroscientist Christian Keysers at the University of Groningen Social Brain Lab and others has shown that people who identify themselves as empathic on self-questionnaires have stronger mirror neuron activity.

The ramifications this research presents for communicating over e-mail are enormous. The visual absence of the person we are in exchange with deprives us of a deep-seated, physical identification with the actions and emotions of others. Marco Iacoboni, the author of *Mirroring People,* says the effects of this are writ large on the Internet: "The rudeness and aggressiveness over the internet—e-mails, blogs, Web forums, etc.—is likely due to the fact that people cannot look into each other's faces and cannot activate mirror neurons, thus cannot activate a very basic process of empathy for other human fellows."

Streams of invective trickle down message boards; comment queues of blogs are marinated in snark, with people blasting the host or one another with angry put-downs. Feedback sections of video-broadcasting sites, such as YouTube, are often a study of life without empathy. At the X Games in 2007, the skateboarder Jake Brown survived a horrifying four-story fall while attempting to land a 720—two-spin (rotation)—on a 293-foot half-pipe called the Mega Ramp. Brown completed the 720 but lost his skateboard on the final ramp, falling fifty feet to the ground and landing with such force his shoes exploded off his feet. For four agonizing minutes, he lay unconscious, potentially paralyzed or dying. "Ha ha ha ha ha ha," wrote one viewer in the comment queue of a YouTube posting of the video. "HIS SHOES POPPED OFF. LOL LOL," wrote another.

Brown ended up walking away from his spill—indeed, in July 2008 he made a successful return to the Mega Ramp—so clearly his hecklers weren't nearly as forceful as his internal willpower. But there are other instances in which disinhibited jeering from the sidelines can cause grave damage. Young kids, such as Megan Meier, who spend more time online than any other group, are in the line of fire of Internet-related disinhibition. A recent study conducted at a middle school in the United States revealed that 17 percent of the student body had experienced some form of cyberbullying, whether it was hostile and threatening e-mails, demeaning posts on Facebook or

MySpace, or videos or pictures posted on YouTube without their permission.

Children and teenagers, whose prefrontal cortexes are still developing, face increasing risks over the Internet, since they are just beginning to learn their inhibitions. "During adolescence there is a developmental lag," Goleman has written, "with teenagers having fragile inhibitory capacities, but fully ripe emotional impulsivity." This leads to flaming and even harassment of teachers. In 2007, Danielle McGuire, a teacher at the New York prep school Horace Mann, discovered that some students, who happened to be children of school trustees, had put up a page on Facebook entitled "McGuire Survivors 2006," portraying her as a witch and a liberal brainwasher. When she asked the school to deal with the situation, she was shocked to discover that the trustees wanted her disciplined for accessing their children's Facebook pages. After a schoolwide disruption, the students were given a slap on the wrist and McGuire was later told the school would not be renewing her teaching contract.

Disinhibition also increasingly leads to sexual bravado. It used to be that teenagers passed notes in class; now, it seems, many of them are e-mailing or texting naked photos of themselves—or others—back and forth. In Santa Fe, Texas, school administrators confiscated dozens of mobile phones after naked pictures of two junior high girls began passing from inbox to inbox. The girls had sent their pictures to their boyfriends, who, like the boyfriend of Claire, who was made infamous for her joke about oral sex, passed them along. In Wisconsin, a seventeen-year-old was charged with child pornography after he posted naked photos of his sixteen-year-old ex-girlfriend on MySpace after she broke up with him. In some cases, naked photos of teenagers have wended their way back to parents.

There's an irony to this state of existence. The computer and e-mail were sold to us as tools of liberation, but they have actually inhibited our ability to conduct our lives mindfully, with the deliberation and consideration that are the hallmark of true agency. We react impulsively, quickly, and must face the consequences later. Our minds, augmented now by the largest, most usable database in the world, are hampered in basic functions such as showing kindness, restraint, and empathy. Digital believers will say that this is just the messiness of true democracy, that we all need to have thicker skins—that there are downsides to all change and the only thing we can count on is change, so adapt or be de-evolved from society.

But if we want to truly have power as individuals, we will preserve the right to push back on this electronic environment that has become such a key component of our day-to-day lives—to tinker with it and, if that doesn't work, resist its basic assumptions as best we can. In coming decades, we're going to have to think hard about whether we want to challenge the urgency, ubiquity, and Wild West quality of electronic communication—because doing so might mean shedding some of the trappings of this newly augmented, free-floating idea of ourselves in order to return to a life where things go a little more slowly.

5

DAWN OF THE MACHINES

In the ongoing television drama *Star Trek: The Next Generation,* one of the creepiest invasion threats comes from a race of cybernetics-enhanced aliens called the Borg. Zipped into bodysuits crenellated with wires and exposed electronics, their headsets gouging deep staplelike grooves into their humanoid flesh, they make a gruesome spectacle of a constantly connected life-form. The Borg—there is no singular—do not operate as individuals but as a hive, their minds plugged into a collective consciousness that they experience in their heads as thousands of voices speaking all at once. The Borg's goal is perfection, and they achieve it by adapting the biological and technological innovations of other species. "You will be assimilated," they state matter-of-factly upon encountering crew ships. "Resistance is futile."

It's hard to find a more potent metaphor for the dangers of the man-machine melding that we have experienced in the last fifty years. Science fiction may not always predict the future, but it is often a brilliant countermythology—a visible cultural symptom—of our prevailing anxieties. Is this the direction in which we are going or how we feel now? A collective society all talking inside one another's heads, in search of perfection, constantly plugged in? It's an extreme example, perhaps, but it's important for us to step back to look at the social implications of our ever-proliferating, ever-accelerating forms of communication technology. It's a task that is harder than ever, given how e-mail has pulverized our days into bite-sized moments of attention.

In fact, one doesn't have to reach forward into science fiction to examine the ramifications—and extremes—of our plugged-in world. We can simply turn back the clock a hundred years to the technology that began this journey to our hyperconnected now—the telegraph—and the people who operated it. Indeed, the creation of the Internet and the explosion of e-mail as a communication tool have turned everyday office workers into the telegraph operators of the twenty-first century.

> Speed was valued above all else: the fastest operators were known as bonus men, because a bonus was offered to operators who could exceed the normal quota for sending and receiving messages. So-called first class operators could handle about sixty messages an hour—a rate of twenty-five to thirty words per minute—but the bonus men could handle even more without a loss in accuracy, sometimes reaching speeds of forty words per minute or more.

One hopes we never get to the point that sending sixty e-mails per hour is normal, but it's worth remembering that telegraph operators were merely *transcribing* messages, they weren't creating them. Still, at two hundred e-mails a day and growing, the average office worker isn't trailing far behind this pace, and the cognitive and emotional juggling required to maintain this rhythm is leading us toward a situation that many telegraph operators experienced: burnout. Telegraph operators regularly lost their tempers; they changed jobs a lot; and some even had breakdowns due to stress. New communication technology is supposed to—and constantly promises to—make our lives easier, but the prevalence of e-mail and its burdens is sending us in the same direction as those men and women who, a hundred years ago, lived out on the electronic frontier.

According to the social psychologist Christina Maslach, who created the first diagnostic inventory of burnout, there are three trademark symptoms: emotional exhaustion, depersonalization, and a reduced sense of individual accomplishment. "How much time do you spend sorting through e-mail?" she asks in *The Truth About Burnout.* "The power of technology is paid for in both time and money." It's also paid for in workers' health.

This sort of workplace stress cost America $300 billion in 2004. That same year, it cost England 13 million working days. In one English study of thirty thousand workers, mental health problems that could be traced back to stress were the second leading cause of missed work after muscle-related issues such as bad backs. It's an issue around the world. Another survey of 115,000 people in 33 countries discovered that excessive work hours and expectations had made work a major cause of health problems. Thirteen percent of the respondents had trouble sleeping at night due to workday concerns. Forty percent of the people who responded said that taking sick days made them feel guilty.

The fact that this rapid speeding up of our jobs is occurring at a stage of capitalism's evolution toward a hyperinterconnected global marketplace where corporations can leverage employees against competing job applicants in lower-wage markets and the increasing use of workers in "nontraditional" work arrangements—such as freelancers and part-timers, many without health insurance or other benefits—makes resisting the process even harder.

The Internet bubble has long since burst, along with the

housing bubble, but our culture has yet to shed its hyperventilated business expectations. Entrepreneurs and workers who can't deliver are failures. "If you work in the Internet business, you're a 25-year-old with a $30 million initial public offering (IPO)," wrote Bill Lessard and Steve Baldwin in 2000. "Anything less means you're an abject loser." Companies have constructed their business model around perennial double-digit growth. Getting this growth requires some draconian measures. In March 2008, Jason Calacanis, the CEO of Mahalo.com, drew waves of criticism when he posted "How to Save Money Running a Startup" on his blog. The list included holding meetings over lunch, buying employees computers so they can work at home and after hours, investing in a good coffee machine, and most controversially, "fire people who are not workaholics."

The expectations and concomitant stress trickle down to people who do not even work in the tech industry, thanks to cultural worship of their success and the groups of people who make money by investing in it through hedge funds, smart stock tips, or just plain dumb luck. In Jonathan Franzen's novel *The Corrections,* Gary, the banker brother of the book's protagonist, Chip, cannot help but feel he somehow missed the boat, that he is behind the times, and that that makes him not just unrich but uncool. "All around him, millions of newly minted American millionaires were engaged in the identical pursuit of feeling extraordinary—of buying the perfect Victorian, of skiing the virgin slope, of knowing the chef personally, of locating the beach that had no footprints. There were further tens of millions of young Americans who didn't have money but were nonetheless chasing the Perfect Cool."

One group of workers who are feeling this strain more than most—aside from the people who have actually lost their jobs—are bloggers, since some of them are tasked with staying on top of the continuous news and product cycle. In 2008, two technology bloggers may even have blogged themselves to death. Russell Shaw, a contributor to ZDNET, and Marc Orchant, another U.S. tech blogger, died as a result of massive coronaries on the job. The pressures of this environment are unbearable. "There's always a new mobile phone whose clunkiness requires dissection, a new security flaw in Windows Vista, or a new USB drive shaped to look like a piece of fruit," wrote Peter Robinson in *The Guardian*. "Keeping on top of it all is an almost impossible task, but people try, and they burn out, knowing that if they sleep, they're scooped."

As e-mail use grows, the stresses of working at this frantic pace will only compound, becoming an ever-stronger feedback loop. We may not quite e-mail ourselves to death, but on heavy days it can feel as if we're getting close to it. A society of people living constantly in this frame of mind does not make for a pleasant place. Impulse gratification is highly catered to in the Western world—it's what keeps the capitalist market running, after all. Most of our purely biological and social needs are simple; we need to create other needs in order to have a reason for purchasing something, for buying based on brand rather than on quality, for believing—as advertisements tell us—that a car or a chair or a pair of jeans will make us a different person, as glamorous as the one in the picture.

By parceling our days into smaller and smaller units, by giving us the impression that we can reach all people, at all times, e-mail is helping to put this cycle of overworking and impatient desire for gratification into hyperdrive. We work to live, the saying goes, but when work takes everything, what's the point? Changing gears at the end of the day only reminds us how much work is taking from us—all while advertisements tell us the sky is the limit. It's almost easier to keep working. And people do.

A 2000 study showed that 40 percent of people work overtime at least once a week—and people who earned between $30,000 and $60,000 were three times as likely as people making less than $30,000 to bring work home. It's almost considered chic to be a workaholic. In early 2008, the pop star Madonna, who had made herself famous as the material girl, admitted to *Elle* magazine that she and her husband, Guy Ritchie, were so busy that *both of them* slept with their BlackBerry under their pillows. "It's not unromantic," she protested. "It's practical. I'm sure loads of couples have their BlackBerrys in bed with them. I often wake up in the middle of the night and remember that I've forgotten something, so I jump up and make notes." News of the couple's impending divorce began trickling out in the summer of 2008.

New technology tools from broadband access to mobile phones to mobile Internet have indeed made work more convenient. The new flexible workplace allows people to skip a commute when traffic snarls, work remotely, schedule their work around their life. "It's the proverbial blessing and curse," said Douglas M. Steenland, the CEO of Northwest Airlines, of his BlackBerry in 2005. "It's a blessing because it liberates you from the office. It's a curse because there's no escape." The downside to this existence, which many non-CEOs now share, is that we never totally shut off and live in the present with our spouses or friends or family. "Blending is a bland term that disguises what is really a new mind-body problem," says Professor Arlie Russell Hochschild, a sociologist at Berkeley. "For some people, like your neighbors and sometimes your children, your body is there but your mind is not. And for others, like your workmates, your mind is there but your body is not. Sometimes it's nice to have both your mind and your body present."

This state of living on parallel tracks has always been part of married life—the challenge always being finding moments to connect. As e-mail went mobile and became domesticated, though, it evolved into a great enabler for people to stay disconnected from one another—an irony given that one of BlackBerry's advertising slogans is "Connect to everything you love." A Berkeley woman whose husband brings his handheld device under the covers described the experience in terms more fitting for an affair. "It's a kind of ménage à trois that I didn't choose, but there it is, every day and night," she said in a *New York Times* article. "There is something about that tap-tap-tap that makes me a little crazy," said another woman, who was considering confining her husband's bedtime e-mailing to business trips.

Spouses are not the only ones neglected when we can't put down our e-mail. A whole generation of children will grow up with ever more distracted parents. In an interview with *The New York Times,* Bruce Mehlman, former assistant secretary of commerce for technology policy, argued that mobile access to e-mail and the Internet allowed him to spend less time at work and more with his kids. He then described having Lego dogfights with his son, one hand on the imaginary plane, the other on his BlackBerry. When he needs a break to clock in, Mehlman made sure to win the "dogfight." "While he rebuilds his plane, I check my e-mail on the BlackBerry," he explained.

There is a small but telling lesson buried in this untouching tableau. The Industrial Revolution, which mechanized production on a scale never before seen in human life, also produced the movement for the eight-hour workday. Men, women, and children couldn't keep up with the durable, mindless muscle of machines. As early as 1817, the Welsh social reformer Robert Owen instituted the slogan "Eight hours work, eight hours recreation, eight hours rest." It took another century, however, for this idea to become law. It wasn't codified in the United States until 1938, when the Fair Labor Standards Act was passed, making it the legal American workday—this in a society that was still climbing out of the Great Depression and into a war. Seventy-five years later, however, our abuse and worship of essentially useful technology have chipped away at this framework. It has readjusted our expectations and, more important, made the company, the corporation and its needs, the dominant context in our lives. And we are going along with it because it makes us feel good, needed, important, connected. We are also being paid less, and to resist would in some cases mean losing our jobs. It is important to note, though, that this is not the only contextual shift e-mail has brought about or accelerated.

It *Is* All About Me

As we spend more and more time online, working at great speed via e-mail on behalf of our employers, it is only natural for us to try and bring the Internet back to ourselves. One of the great paradoxes about e-mail is that although it is created, driven, and indelibly marked by ourselves, heavy use of it can leave you feeling emptied out, voided, fractured into a million bits and quips, yet somehow obliterated. The cultural critic Fredric Jameson's comments on postmodern architecture could apply well here: "A constant busyness gives the feeling that emptiness is here absolutely packed, that is an element within which you yourself are immersed, without any of that distance that formerly enabled the perception of perspective or volume. . . . You are in this hyperspace up to your eyes and body."

In the past couple years, several tools have developed to help us feel—in this new work and living environment—whole again. Social networking Web sites such as MySpace and Facebook are at the forefront of this movement and have become hugely popular for this reason. Facebook was launched on the campus of Harvard University by a sophomore, Mark Zuckerberg, in February 2004 as a way to help his classmates connect online. Within a month, half of the university's undergraduates had signed on; a month later it was expanded to several other Ivy League universities and thereafter to colleges and high schools. Soon everyone wanted to join in. More and more users were over thirty years old, but Facebook's greatest support continued to be among college students. A 2006 survey found that Facebook was the second-most popular thing among college students after the iPod; it tied for second with beer. In 2006 it had 9 million registered users. By the summer of 2009, that number had rocketed to 250 million.

Ticker tape, which was developed from the telegram for the purpose of business and urgent news, has found a twenty-first-century use in transporting the news about ourselves on sites such as Facebook, which encourages you to broadcast what "you are doing" to your friends, or Twitter, which allows you to post short messages to a "channel" that can be accessed (and commented upon) by handheld or computer. In fact, given how much time many people spend on these sites—or interacting through e-mail—it's fair to say that *we are now the news*. It's a fascinating reversal from just seventy years ago. In John Dos Passos's epic *U.S.A.* trilogy, bursts of ticker-tape news interrupt the story, putting the lives of his characters into the shadow of an ever more rapid present: these snippets of news also continuously called into question what was the defining context of a life—the swell of history it rides upon or the thoughts of the travelers in the moment? "I wish I was hard enough so that I didn't give a damn about anything," says one character. "When history's walking on all our faces is no time for pretty sentiments."

A lot has changed in postwar America and around the globe—

not just economic circumstances—to make this not so. Advertising promoted the idea of individuality to the point that people's desire to appear special, different, and other actually made them seem more like a crowd, a trend Hal Niedzviecki has described hilariously in *Hello, I'm Special,* such as the generation of disaffected teens who bought carefully distressed, grubby-looking clothes at Urban Outfitters and all wound up looking the same. Also, as Jean Twenge argues in *Generation Me,* an entire generation of children was raised to believe that self-esteem was the most important thing in their well-being. This led them to feel more entitled than ever and emptier when their high expectations were not fulfilled. In this way, around the Western world but especially in America, one of the most coddled generations ever created is more miserable than can be imagined.

One offshoot of this new narcissism is "egosurfing," in which one searches the Internet for information about oneself. A company—egoSurf—was created especially to do this. Google, however, quickly took over the market; you can now simply put a Google alert on your name, and any time you are mentioned on the Internet an e-mail will be sent to your inbox. Even the famous—perhaps especially the famous—are afflicted by it. In Tim Parks's 2006 novel *Cleaver,* a disgraced news presenter retreats into the Austrian Alps to lick his wounds. Surrounded by mountain vistas and the ticking silence of a lush remove, he cannot help ducking into an Internet café to Google himself.

As Cleaver discovers, once we project our self out into the world and begin tracking it on the Internet, no amount of feedback is enough. The pit of identity vertigo is bottomless. "Friend confirmations come in every minute," wrote Brian Palmer, a student in Pittsburgh. "How can I not click onto Facebook and see if someone new has listed me as a friend? But Facebook doesn't make me feel like I have friends. Friends aren't supposed to let you sink deeper into an addiction." No amount of virtual connectivity will ever suffice, either, because face-to-face interaction provides things that online conversation can never deliver: touch, the complex emotional valences of expression and smell, inflection or tone of voice, the awkward but essential jaggedness of being present in the world. "There is a kind of famine of warm interpersonal relations, of easy-to-reach neighbors, of encircling, inclusive memberships, and of solid family life," says the social scientist Robert Lane. "I have never checked my e-mail more obsessively in my life," wrote Twenge of her experience in online dating.

The Loss of Public Space in the Physical World

> *The middle distance fell away, so the grids (from small to large) that had supported the middle distance fell into disuse and ceased to be understandable. Two grids remained . . . there was a national life—a shimmer of national life—and intimate life. The distance between these two was very great. The distance was very frightening. People did not want to measure it. People began to lose a sense of what distance was and of what the usefulness of distance might be.*
>
> —George W. S. Trow, *Within the Context of No Context*

In 1980, alarmed at what he felt television was doing to American life, George W. S. Trow published one of the most prophetic essays about the direction of what was then called the Media Age. Trow believed we had been lured from smaller, tangible groups, such as women's clubs and bowling leagues, out into the false camaraderie of television programs. Tuning in to the tube, we instantly exchange a virtual world for the real one. We have friends because we can tune in to the crew of *CSI: Miami* most nights. Newspapers, which reported the world, gave way to television news pitched to viewers based on demographics, a contextual shift that eroded people's sense of history on both a large and a small scale. The vacuum created by the loss of what Trow called "middle distance" institutions was quickly filled. "It is in this space," Trow wrote, "that celebrities dance."

The breaking down of institutions that Trow was commenting upon, places where face-to-face interaction was at a premium, has only been speeded up by the Internet. These days you don't have to leave the house. In 2000, a twenty-six-year-old north Dallas man pulled one of the more successful publicity stunts of recent memory by legally changing his name to DotComGuy and vowing not to leave his home for a year. The move earned him scores of sponsors—United Parcel Service, Gateway, a gym that sent a trainer to the house so he wouldn't get too pudgy—and millions of viewers to his Web site, which broadcast his life via live webcams. People spent 2 million minutes watching him in the first four days of his sojourn.

In the wake of the Y2K craze, when suddenly the matrix of computer technology we were becoming dependent upon seemed to have exposed us to a fatal flaw, DotComGuy was a great advertisement for the pleasures of the Internet. Apparently, many people bought into this experience. Online shopping took off, growing from the inception of the Web to a rate of 25 percent per year before slowing—it is the second-most common thing Internet users do after e-mailing—but it's been at the expense of the real-world commons, that is, the place where one can interact face-to-face in real time, unmediated. Indeed, stores like Barnesandnoble.com began to offer the option to purchase a book online but pick it up in the store for customers who actually liked the chance to be in a physical retail space. "It's not like you go onto Amazon and think, 'I'm a little depressed. I'll go onto this site and get transported,'" said Nancy F. Koehn, a Harvard Business School professor who studies consumer habits. But it is exactly this instinct that pulls us into a beckoning bookstore.

For those who cherish the corner store, the migration of such a huge range of business onto the Internet couldn't have occurred at a worse time, because these real-world, middle-distance commons—places where you could interact without the emotional strain of carrying a long conversation—as Trow noted, were already suffering. When William Penn was governor of Pennsylvania, the mail service's delivery times and routes were simply posted on the meetinghouse door in Philadelphia. This community collective has dwindled drastically ever since. "Within the twentieth-century city of housing," Joseph Rykwert wrote, "the identifiable places of meeting have been drastically reduced." Retail outlets, fast-food restaurants, chain stores "convey the message that space has been standardized, that its inflections and associations have been ironed out."

It's hard to blame us, then, for retreating into a virtual space, withdrawing more and more into the window of our computer screens to do and say things that are difficult in real life. In making this compromise, however, we are buying into a philosophy of space—an ecology of space that is designed and determined by the systems that drive the Internet. As Mark Rothko has written about art, "If one understands, or has the sensibility to live in, the particular kind of space to which a painting is committed, then he has obtained the most comprehensive statement of the artist's attitude toward reality." If we agree with cultural critic Steven Johnson that an interface is art, the question remains: What are the assumptions and worldview of the screen? For starters, one could say a computer screen leads us to believe that all the world is available from our fingertips; that there are no limits; that there is no time but now. That real space doesn't matter.

And so, by tying ourselves to this machine, we make a trade: virtual interaction for physical togetherness. The millions of people who simultaneously make this bargain one click at a time have had an enormous effect on the institution that brought us together to begin with: mail. In 2008, with the post office running a deficit of £4 million a week, the British government announced a plan of post office closures that would affect 2,500 branches. Many of them were small, irreplaceable meeting places for their communities. It was a particular blow to the elderly, who pick up their pension and benefit checks at the post office, and the infirm. "We have to remember that there are many who don't have cars, not to mention many who don't have home computers," wrote a man in Shropshire. "And even though there have been improvements to public transport, in rural areas buses don't always run as frequently as people need." One of the oldest post offices in the country, operated by the earl of Leicester, was closed. Scores of people complained. One community even made a film in protest. Yet the cuts went through.

This erosion of face-to-face interaction is taking place all over, anywhere people traditionally gathered in public. Instead of going to a movie theater, people can download a film over the Internet; instead of walking into a bookstore to browse, customers simply navigate to an online store; shoppers pressed for time do not visit a local grocer, they buy their produce on FreshDirect or a similar site; pickup bars are losing out to Internet dating sites, auction houses to eBay; casinos don't have to go offshore anymore, as people can go online to play poker; banks' queues are diminishing as more and more customers go online to perform simple transactions or pay bills; garage sales are going away as people hock their old goods on Amazon.com, eBay, and craigslist; even doctors' offices can be avoided. About the only thing you cannot do from home, at this point, is vote.

This migration into the virtual space is being felt most strongly by people who grew up without it. But a new generation is being raised to take it for granted. Schools and universities, which used to be designed around quadrangles in order to facilitate meetings, have become as eerily silent as offices now that students gather virtually, updating their Facebook pages at a frenzied rate rather than actually talking to one another. In 2000, a *New York Times* reporter visited the campus of Mount Saint Mary College, which had installed campuswide wireless Internet access, then a novelty. The new connectivity was an enormous convenience to many on campus, but it had quickly created a dependency. People were on it all the time. "I've never seen a student walking around holding their laptop out to listen to music," said the network manager, tuning the Internet radio to a new station. "But I'm sure that's not far off." The service was absorbed so quickly, the reporter wrote, that students and faculty "developed an automatic reflex to go to the Web for information, no matter where they are."

As cheap, easily accessible clearinghouses for news that affects people in a particular geographic region, newspapers—another form of public space—are also in trouble, especially in the United States and the United Kingdom, where 150 years ago the telegraph gave them a new life and expanded the sense of how far empathy's tether should reach. Relaxed FCC laws have allowed corporations to buy up and consolidate more media than ever—but it's increasingly difficult for newspapers to produce the double-digit growth that made them such attractive parts of a media empire. Newspapers typically made 80 percent of their money from advertising and 20 percent off circulation. Online sites such as craigslist have slowly eaten away at their advertising market, and more and more people are reading the news online. Newspapers are still expected to report and explain the news; they just can't make money distributing it the same way they used to.

E-mail has played a role in this development. Our inboxes have usurped the morning paper as a shaping context; many of us check it before we even glance at the news, let alone brew that first cup of coffee, making our e-mail (and by extension ourselves) the most important information—the shaping context—of the day. This is an important development. From dawn to dusk, e-mail has become a kind of rolling to-do list that, as more and more information is provided to us electronically—from sales at our local department store to news of a friend's birthday to important deadlines for work—stretches across all aspects of our life. If this is the first stream of information we dip into in the morning, we begin our days with a contracting sense of the world, rather than an expanding one. The possibility of serendipity, of learning something we don't already know or at least think we may want to know, diminishes.

This isn't to say that we cannot get the news delivered to our virtual inboxes. It's just a very different notion of what news means. Google, for instance, has made it possible to have news delivered to your e-mail address through Google news alerts, which appear embedded in an e-mail as a series of hyperlinks. Simply type in the list of subjects you would like to track, and it will e-mail you alerts as many times a day as you can handle. Markos Moulitsas, the founder of the political blog Daily Kos, which gets upward of a half million visitors per day, told *GQ* he now gets all his news via Google alerts.

Shopping for news this way, however, puts a modern-day reader into the same dilemma that the Michigan newspaper mentioned earlier was in with the telegraph many years ago. Things like Google alerts, RSS feeds sent out by e-mail, or links on your Yahoo! home page often remove local factors—such as your newspaper—as your defining context, your window onto the world, because it depends entirely upon what is written about and reported over the Web. One would think this would present an opportunity for local newspapers, but as more and more of them have been bought up by corporations, forced to use syndicated copy that comes from someplace else, reduced to a skeleton staff of writers and reporters, these papers, which would have stood to benefit from the overload of information provided by the Web, are actually losing out.

Not surprisingly, then, in the United States, the past few years have been bad ones for local newspapers—that, to be fair, were slower to pick up on the possibilities the Web held for reporting and that had fewer resources to market and promote themselves—and very good ones for papers pitched to a national audience, such as *USA Today, The Wall Street Journal,* and *The New York Times.* But even these giants face a larger contextual shift away from news, toward a point-and-hunt attitude to information.

Google is a powerful cultural force, but it alone is not responsible for this trend; the customer has shaped and designed the news without knowing it ever since demographics and ratings became a large part of news reporting, especially for broadcast news. The Web and tools such as e-mail have simply put us more firmly into the role of consumer: What do we want? *The New York Times* has a feature that allows a reader to customize his or her news home page, making their news site into a mash-up of headlines pulled from a series of news sites—from *The New York*

Review of Books to *Sports Illustrated* to the Huffington Post—on a page you design and lay out yourself. What the newspaper itself considers important and newsworthy and where it ought to be placed on your attention span doesn't matter.

The struggles faced by regional newspapers in the United States—and now in the United Kingdom—does not affect simply the owners and reporters and paperboys; it affects what citizens can know. National news agencies do not have the resources to cover all the news that is fit to print; local papers have been essential in filling in the gaps. They have exposed corruption, graft, abuse, murders, and rapes—defended the weak from the powerful—in a way that never required an economies-of-scale justification. For a story to be reported in these markets, it need only affect the people in the circulation areas. If local news sources—especially newspapers, which form the backbone not just of radio and television but of many of the blogs that have arisen to fill the gaps—are allowed to die on the vine, people in those communities will be deprived of an essential watchdog of the commons. Fighting this trend will be an uphill battle. Many of these newspapers are now owned by people who do not even live in the community. Moreover, we have entered a climate in which our frame of reference has shrunk to the smallest aperture ever.

Why Is It So Hard to Read Anymore?

The nature of screens—and how we work on, over, and through them onto the Internet—has effected a huge epistemological shift that goes beyond merely writing our way into existence. It's changing how we read and what we read, just as the telegram and organized mail did before it. As Nicholas Carr wrote in his essay on Google, "It is clear that users are not reading online in the traditional sense; indeed there are signs that new forms of 'reading' are emerging as users 'power browse' horizontally through titles, contents pages and abstracts going for quick wins. It almost seems that they go online to avoid reading in the traditional sense."

Some of these changes stem from the interface itself. Newspaper and magazine articles have become shorter, breaks longer, and text bigger to accommodate readers' fractured attention span. We ourselves create this condition, though, by how much time we spend working in word processing programs and on e-mail. "E-mailers tend—there being no space constraints—to insert a line of space between paragraphs," writes the humorist and language columnist Roy Blount, Jr., in an e-mail. "If readers get to where they can't tolerate paragraphs without space between, they will develop an even greater resistance to print; or print will have to put space between paragraphs, which will eat up more paper, make books bulkier, leach even more substance out of newspapers and magazines: contribute further to the decline of print."

Empirical evidence is flooding in regarding the ways that screen-based reading, which has grown from e-mail, is changing the way we read generally. Eye-tracking studies have shown that people increasingly tend to leapfrog over long blocks of text. We need bullet points, bold text, short sentences, explanatory subheads, and speedy text. People skim and scan rather than rummage down into the belly of the beast. Online readers are "selfish, lazy, and ruthless," said Jakob Nielsen, a usability engineer, in an amusing article by Slate's Michael Agger. Instead of explaining concepts in a text, Nielsen advised putting in hyperlinks to other articles where readers can pick up other concepts. Even the most usability-enhanced article and layout, however, can lose out on the Web. "[Spontaneous reading pleasure] can be achieved," Agger writes, "but the environment works against you. Read a nice sentence, get dinged by IM, never return to the story again."

More and more people—increasingly young ones—are doing their reading in this environment. In 2008, a Swarthmore College English major read the entirety of Henry James's *The Golden Bowl* on his BlackBerry; the small screen and being interrupted by e-mails and IMs didn't bother him (perhaps this says more about James than the student). He is not alone, if not in reading James on a screen, then in reading a book on a handheld. The company Fictionwise, which provides e-books for PalmPilots, iPhones, and other handhelds, has a staggering inventory of 53,000 titles, equaling 971.6 billion words, used by more than half a million users.

In July 2008, the *New York Times* reporter Motoko Rich went to Ohio and reported on how a fifteen-year-old consumes her nightly media diet. "Nadia checks her e-mail and peruses myyearbook.com, a social networking site, reading messages or posting updates on her mood. She searches for music videos on YouTube and logs onto Gaia Online, a role-playing site where members fashion alternate identities as cutesy cartoon characters. But she spends most of her time on quizilla.com or fanfiction .net, reading and commenting on stories written by other users and based on books, television shows or movies." Nadia's mother wasn't necessarily worried: "I'm just pleased that she reads something anymore."

It's not a huge surprise, at least in the United States, that there has been a huge dropoff in how many Americans, especially young Americans, read for fun. Less than one-third of thirteen-year-olds are daily readers, a 14 percent decline from twenty years ago; on an average day, Americans aged fifteen to twenty-four spend two hours watching television and just seven minutes reading. As a result, reading test scores are plummeting, even among well-educated adults. From 1992 to 2003, the percentage of adults with graduate school educational experience who tested as proficient in prose reading dropped by ten points. Critics argue that these results discount reading done online, but so far the benefits of that reading are hard to discern. "What we are losing in this country and presumably around the world is the sustained, focused, linear attention developed by reading," said Dana Gioia, a former chairman of the National Endowment for the Arts. "I would believe people who tell me that the Internet develops reading if I did not see such a universal decline in reading ability and reading comprehension on virtually all tests."

Rich and the NEA both point out that "reading comprehension" remains one of the top skills in demand for well-paying jobs—even if it's not clear what kind of reading will be the most valued (scanning or scrolling, or sustained) once the employee starts. If the research on multitasking is any guide, though, and if several centuries of liberal arts education have proven anything, the ability to think clearly and critically and develop an argument comes from reading in a focused manner. These skills are important because they enable employees to step back from an atmosphere of frenzy and make sense in a busy, nearly chaotic environment. If all companies want, though, is worker bees who will simply type till they drop and badger one another into a state of overload, a new generation of inveterate multitaskaholics might be just what they get. If that's the case, workplace productivity isn't the only thing that will suffer.

History in the Wash

The speed we must work at in order simply to manage our correspondence has extended the disposable culture that has ruined so many parts of our globe to our own correspondence. We delete e-mails to clear space—and often our machine does it for us. Computer crashes, for those who do not use Web-based e-mail, routinely wipe out large correspondence files. Given that we are having conversations via e-mail that once occurred over the telephone—which no one but the most paranoid recorded—these eclipses are in keeping with how much was lost to time, anyway. Many people are sending e-mails rather than letters, though, and letters were kept, treasured, archived. Not preserving such correspondence will lead to a terrible loss.

Family histories depend to a large degree on letters; they create a sense of continuity, tell us things loved ones could not share—or had forgotten—in their lifetime, and are one of our last remaining physical contacts with the presence of our ancestors. There is nothing quite as posthumously intimate as handwriting. It's why we rummage. Letters that emerge from the dust of attics and lockboxes are like time capsules; they preserve the concerns and worries and passions of our loved ones at the moment they were written. Novelists know the power of these words. Some of the earliest works of fiction were epistolary tales. In Samuel Richardson's *Clarissa* and Fyodor Dostoevsky's *Poor Folks,* we follow the story by reading the letters of characters. We know things about them that are so private they can be put only into letters.

Even if e-mails are preserved, though, they may not tell us as much as letters. The way the medium conditions us to write will also pose a problem for future historians. "In our age when diaries are no longer regularly kept, when letters are rare between lovers and friends, the only hope for future historians will be email," says Doris Kearns Goodwin, the biographer of Abraham Lincoln, Lyndon Johnson, and the Kennedys. "Yet, compared to handwritten letters in the old days, in which the writers often poured their most intimate thoughts, the more staccato form of e-mail will be nowhere near as valuable. There is also the fear that when a couple breaks up, for instance, they will simply delete whole files, which was less likely when handwritten letters were retained in old boxes and stored in attics for generations."

We are making history faster than we can preserve it. In *The Future of the Past,* Alexander Stille described how the race to preserve archives and collections often overlooked the fact that new methods of preservation were actually inferior to what they were replacing, like that old piece of clay that has housed a love poem for four thousand years and running. "In fact, there appears to be a direct relationship between the newness of technology and its fragility," he writes. "The clay tablets that record the laws of ancient Sumer are still on display in museums around the world. Many medieval illuminated manuscripts written on animal parchment still look as if they were painted and copied yesterday. Paper correspondence from the Renaissance is faded but still in good condition while books printed on modern acidic paper are already turning to dust. Black-and-white photographs become unstable within forty or fifty years. Videotapes deteriorate much more quickly than does traditional movie film—generally lasting about twenty years. And the latest generation of digital storage tape is considered to be safe for about ten years, after which it should be copied to avoid loss of data."

These issues are not just important for archives; they are essential to government, for gaps in the record can enable an administration to effectively rewrite history. Following President Nixon's raid on White House phone tapes, the U.S. Congress passed the Presidential Records Act, mandating the preservation of all presidential records. But the age of e-mail has created new loopholes—it's easier than ever to lose, delete, or write over existing records. In the 1980s, the administration of President Ronald Reagan sought to destroy records from an early electronic messaging system; in the end, it was blocked by a court. The Clinton administration also became ensnarled in a legal battle relating to its failures to preserve electronic records. In the end, the White House e-mails had to be reconstructed at a cost of several million dollars.

No administration lost quite as much mail as the George W. Bush administration, though—a double standard if ever there was one, given that it also peered into e-mail more invasively than any other U.S. administration to date. In 2007, it was discovered that e-mails of high-level staffers involved in the dismissal of eight U.S. attorneys had gone missing. It was also discovered that the Bush administration had encouraged many of its staffers to use a domain provided by the Republican National Committee (RNC), "in some instances . . . specifically to avoid creating a record of the communications," according to the chairman of the House Committee on Oversight and Government Reform, Henry Waxman. The RNC had no e-mails archived for fifty-one of the eighty-eight officials who had such accounts, which were apparently used to discuss policy. Karl Rove, the president's notoriously wily adviser, claimed to have used the RNC domain for 95 percent of his e-mail.

This was not the first or last time that White House e-mail turned up a blank during George W. Bush's presidency. When special prosecutor Patrick Fitzgerald was investigating the outing of CIA agent Valerie Plame, he discovered that e-mails were not always backed up. In May 2008, the Bush administration admitted in court that it had lost three months of e-mail from the initial months of the Iraq War thanks to a system upgrade. Facing a lawsuit by the National Security Archive, a nonprofit group that focuses on uncovering classified documents, the administration also said the earliest date on e-mails was May 23, 2003, the day the United Nations gave official approval for the occupation of Iraq. All told, between 2003 and 2005, the Executive Office of the President lost more than 5 million e-mails.

The Explosion of Now

What we are experiencing in the twenty-first century is a shift in the space-time continuum as radical as the one brought about by the creation of simultaneity by the telegraph and the institution of standardized time. Prior to the development of e-mail and the use of e-mail before and after work to keep up with matters not resolved during the workday, one had to actually call a coworker at home to keep a project churning, and the invasiveness of making that phone call helped to reinforce an essential boundary between public and private space, between work and leisure time, between family priorities and business priorities.

We have a new now, a now that doesn't care about time zones or distance, a now that is muscularly, aggressively rearranging our lives and circadian rhythms because, unlike people the world over in the nineteenth and early twentieth centuries, we are plugged into the network, which connects us more than ever. In the past, very ambitious businessmen, world leaders, and telegraph operators were obliged to field messages coming in over the wireless at night. Now that requirement has expanded massively—we need that update from the Sydney office, the markets are open in Hong Kong, checking in gives us a jump on the day, erases the lag time that the scope of the world's physical dimensions has until recently made all but insurmountable. The corporation never sleeps.

Globalization has been with us for centuries now, but the use of technology such as mobile phones, voice mail, and e-mail has physically corralled us into the same sleepless pool. In this environment, those who can go without are demigods, paragons of hard work and mind over matter. Bill Clinton, Martha Stewart, and Margaret Thatcher were and are renowned for their ability to push on into the wee hours. "Sleeping as little as possible is viewed as a badge of honor here," says Dr. Eve Van Cauter, a sleep researcher at the University of Chicago. An NPR story in 2008 brought forward one of the classic examples of a power broker's attitude to snoozing: In the film *Thank You for Smoking,* the main character, Nick, a public relations flack for the cigarette industry, receives a phone call from a Hollywood poobah late at night. "Do you know what time it is in Tokyo?" the Hollywood man quips, saying yes, of course he's in the office. "It's the future!" "When do you sleep?" Nick asks. "Sunday!" the mogul replies.

Trying to keep up, the rest of us suffer. Insomnia and chronic sleep deficit have become something of an epidemic in the United States. Americans are sleeping on average one less hour than they did just twenty years ago, and our caffeinated, constantly connected infotainment culture is to blame. "All of these lifestyle changes are directly impacting not only the number of hours Americans sleep each day, but also when during the twenty-four hours that sleep occurs," says Carl Hunt, the director of the National Center on Sleep Disorders Research in Bethesda, Maryland. And we may be able to work into the wee hours, but it's not efficient—it simply extends the hours we need to work in order to get things done. Among Americans aged eighteen to thirty-four, 50 percent say that daytime sleepiness affects their work. These circadian blips are costing U.S. companies $50 billion a year, according to some estimates.

It's not just our rest that suffers, though. Living in a perpetual now alters our perception and arguably makes it impossible to grasp the dimensions of our lives and our world, since our attention has constantly to be oriented to the task at hand. This means that in the so-called age of information we are constantly making decisions based on a limited, improperly grasped frame of reference—in other words, on faulty information. In *The Dimension of the Present Moment,* the poet and immunologist Miroslav Holub relates a common experiment in which a subject is presented with a brief sound or light signal and is then asked to reproduce it. If the signal lasts for just two seconds, the subject always overestimates its length, creating a reproduction that is longer. If the signal played to the subject is three seconds long, then the reproduction is accurate.

The symbolic and psychological ramifications of this experiment are enormous, especially in an era in which we are constantly required to interact in the present moment—when we have turned an asynchronous tool like e-mail, one that, technically, doesn't require two people to be present in the same moment—into a rapid-fire messaging system that gives us constant and permanent access to everyone we know. E-mail may give us the sense that we are all together now, but science tells us that we are all still living on slightly different senses of time, just like the people living in between major time stops on the American railroads before the dawn of standardized time. "We live permanently removed from and critical of our own past," Holub writes, "permanently removed from and in the hopes of the oncoming future." We are also increasingly cut off from parts of the world that are considered *out there.*

The Digital Divide

The truth remains that it's not just nature that is lost on e-mail; it's the true collective. Marshall McLuhan, the so-called godfather of the wired generation, coined the term "global village" forty years ago, when ARPANET was becoming a reality, yet that term remains a dream, not a description. Many groups remain stubbornly untouched by—or unmoved by—the so-called Internet revolution. The elderly, for instance, do not use the Internet or e-mail much. In the European Union, age and education remain determining factors for whether or not people use e-mail and the Internet. A survey conducted in 2004 revealed that people aged sixteen to twenty-four were three times as likely to use the Internet as those aged fifty-five to seventy-four; the same ratios divided those with lower and higher educational levels. Income also plays a factor. In Canada in 2007 a survey showed that 91 percent of people who made $91,000 or more regularly used the Internet; only 47 percent of those who made $24,000 did so. Between 2007 and 2008, the number of Americans who had broadband connections rose from 47 to 55 percent, but none of this growth came from poor populations, even though broadband costs were nearly 5 percent less than they were in 2005.

It's not hard to see why the poor aren't nearly so connected. Most libraries in the Western world have computer terminals now, but the hours of operation are not convenient for people who work during the day—or work two jobs, as many of the poor must in order to make ends meet. Most workplaces are connected, but that's not true for service jobs—where a great deal of lower-income people work—at fast-food restaurants, or in retail stores. Their employees need to be out on the floor selling or flipping burgers or cleaning out plastic booths cluttered with trash. It's actually against these companies' interest for their workers to have Internet access. The same goes for factory and manual labor jobs. Migrant workers do not have computer labs at their "place" of work; neither do millions of people the world over who work in sweatshops, mines, canneries, and sawmills and on road repair crews and short-order cook lines.

Nations, the renowned historian Benedict Anderson argued, are imagined communities. They function based on the agreed idea that people who live in the same geographic area or occupy a similar group can remain in that group without ever meeting face-to-face. "Yet in the minds of each lives the image of their communion." So a dog walker in Portland, Oregon, and a ditch digger in Portland, Maine, are both Americans—and can both identify as such—and need not see or communicate with each other in order to agree upon that fact.

The Internet is the largest imagined community ever built. But it is also as closed as, if not more so than, the imagined community of nations. For to be online you need money. You need a job that provides a computer, or you must own one; you must possess computer literacy, which isn't always available for people who drop out of school or do not get schooled at all. It also helps if you have people online with whom to write or send messages—so a group's reluctance to go online, say certain populations of seniors, reinforces itself. In many cases, particularly in the world's poorest countries, as much as Internet use and communication would help them, there are far more pressing concerns, such as simply having enough to eat. In 2008, only 5 percent of Africa's population was online.

In this sense, the Internet and e-mail, while designed to make the world ever more connected, reinforces the already existing gaps. Mighty work is being done by several charitable organizations, such as Nicholas Negroponte's foundation One Laptop per Child, to help provide computers and Internet access to more libraries and communities, to get parts of the world wired that still don't even have dial-up. Helping poorer countries and populations get connected is important work, and it should be done. In the meantime, it is important for those of us who are in the Internet's imagined community to remember that this digital utopia doesn't exist for everyone.

The Age of Missing Information

In May 1992, Bill McKibben performed a fascinating experiment. He recorded an entire day of television—all ninety-three channels in Fairfax, Virginia—and compared the information it provided with the information acquired in a day spent atop a mountain in the Adirondacks. It was a straw man context, he admitted, since no one can physically watch ninety-three channels at once, but the comparison provided an illuminating argument for what is missing in the so-called age of information. Many people spend a fifth of their waking life before the machine; what is it telling them?

One key ingredient missing from television, of course, is any sense of the natural world and its constant reminder of human hubris. "Even the dullest farmer quickly learns, for instance, a deep sense of limits," he wrote. "You can't harvest crops successfully until you understand how much can be grown without exhausting the soil, how much rest the land requires, which fields can be safely plowed and which are so erosion-prone that they're best left to some other purpose."

Now that we're plugged into a different screen, the computer screen, for many hours per day, this denaturing has only increased. Many of us who work in offices don't touch a single natural substance all day long, from the plastic keyboard to the mesh fibers on the back of our chair to the faux lacquered tables in conference rooms. These tangible objects are the new machine world, and their message is that human engineering drives the day; our hands, our designs are stamped upon everything. We have forged an environment made for work.

By projecting ourselves into the computer, by imagining it as a surrogate brain, an extension of ourselves, we have bought into a false metaphor that travels with us everywhere our gadgets go. The computer is not a brain any more than our brain is a computer. The computer, for one, is far better at batching and sorting than we are; it can work longer hours. It feels no pain. But we cannot help projecting ourselves into and through it because not to do so would be to recognize how much we live in a world that is outside ourselves—denatured, not human. This paradox puts us into an emotional and physical bind. We have become dependent upon a machine that cannot sense our physical strain and has no intuitive knowledge of our limits, a machine that is immune to sunlight—and in some cases, does not work well under direct light due to the glare. In turn, this relationship has pushed us far beyond our known limits. As Elaine Scarry has written in her treatise *The Body in Pain,* "The act of human creating includes both the creating of the object and the object's re-creating of the human being, and it is only because of the second that the first is undertaken." In other words, it is not the machines that are doing this to us: it's ourselves, because we keep inventing the machines. We have collectively desired this augmentation of our power, our reach, our communication skills. But if we want to change the way we are living and working, we have to acknowledge our limits and the fact that we may have reached terminal velocity a long time ago.

6

MANIFESTO FOR A SLOW COMMUNICATION MOVEMENT

The best hope for emotional maturity, then, appears to lie in recognition of our need for and dependence on people who nevertheless remain separate from ourselves and refuse to submit to our whims. It lies in recognition of others not as projects of our own desires but as independent beings with desires of their own. More broadly, it lies in acceptance of our limits. The world does not exist merely to satisfy our desires; it is a world in which we can find pleasure and meaning, once we understand that others too have a right to these goods.

—Christopher Lasche

In a secret way, we have always wanted to speed up to this point—to accelerate into "that instant," writes Hélène Cixous, "which strikes between two instants, that instant which flies into bits under its own blow, which has neither length, nor duration, only its own shattering brilliance, the shock of the passage from night to light." In the instant we are everywhere and nowhere, the boundaries of space and time do not matter; we will never die. The instant between 0 and 1.

But the boundlessness of the Internet always runs into the hard fact of our animal nature, our physical limits, the dimensions of our cognitive present, the overheated capacity of our minds. "My friend has just had his PC wired for broadband," writes the poet Don Paterson. "I meet him in the café; he looks terrible—his face puffy and pale, his eyes bloodshot. . . . He tells me he is now detained, night and day, in downloading every album he ever owned, lost, desired, or was casually intrigued by; he has now stopped even *listening* to them, and spends his time sleeplessly monitoring a progress bar. . . . He says *it's like all my birthdays have come at once,* by which I can see he means, precisely, that he feels he is going to *die.*"

We will die, that much is certain; and everyone we have ever loved and cared about will die, too, sometimes—heartbreakingly—before us. Being someone else, traveling the world, making new friends gives us a temporary reprieve from this knowledge, which is spared most of the animal kingdom. Busyness numbs the pain of this awareness, but it can never totally submerge it. Given that our days are limited, our hours precious, we have to decide what we want to do, what we want to say, what and who we care about, and how we want to allocate our time to these things within the limits that do not and cannot change. In short, we need to slow down.

Our society does not often tell us this. Progress, since the dawn of the Industrial Age, is supposed to be a linear upward progression; graphs with upward slopes are a good sign. Processing speeds are always getting faster; broadband now makes dial-up seem like traveling by horse and buggy. Growth is eternal. But only two things grow indefinitely or have indefinite growth firmly ensconced at the heart of their being: cancer and the corporation. For everything else, especially in nature, the consuming fires eventually come and force a starting over.

The ultimate form of progress, however, is learning to decide what is working and what is not; and working at this pace, e-mailing at this frantic rate, is pleasing very few of us. It is encroaching on parts of our lives that should be separate or sacred, altering our minds and our ability to know our world, encouraging a further distancing from our bodies and our natures and our communities. We can change this; we *have to* change it. This book has been an attempt to step back from the frenzy and the flurry of the now—the now we have created and the now we have to slowly remove ourselves from—to make this argument. Of course e-mail is good for many things; that has never been in dispute. But we need to learn to use it far more sparingly, with far less dependency, if we are to gain control of our lives.

The reduction of distances has become a strategic reality bearing incalculable economic and political consequences, since it corresponds to the negation of space.

—Paul Virilio

Each new birth of reality, however deformed, can be exploited in its turn.

—George W. S. Trow

The first parameters—the only parameters—for human existence are natural ones. Even our most dazzling cities are embedded into ecologies that scratch stealthily at the steel and brickwork, ready to reclaim the ground upon which skyscrapers and streetlamps are built. Wolves haunt Central Park at night. An empty suburban home, as it crumbles, will be taken over by thousands of living organisms that feed off its timber, dig roots into the floor joists. We are short-term leasers in this world, even if there are concrete beneath our feet and fiberglass roofs above our heads. Were humans to abandon the streets of New York City, they would collapse into the subway tunnels in just two decades.

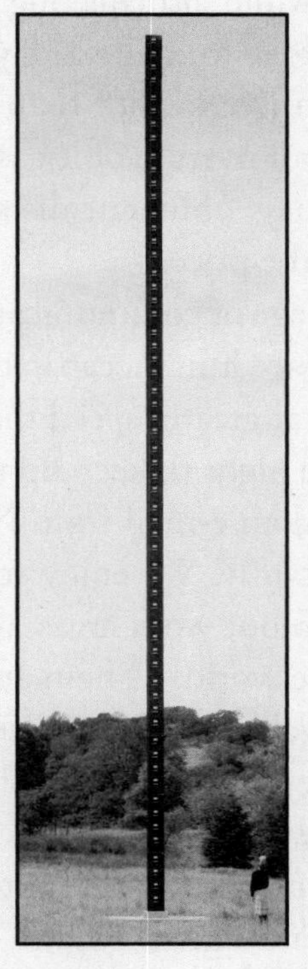

The modern day office: it goes with us everywhere

Technology amplifies human instincts and desires, but it must obey the laws of nature if it is to sustain human life, not destroy it. And so we must remember we are part of nature, too. We may be dependent on machines, but we operate like them at our peril. A diving suit can sustain a swift ascent from 3,000 meters below sea level; the human body inside it cannot. A man who works past the point of exhaustion in a mine will collapse; a machine can keep on digging.

Technology that appears to transcend the limits of the physical world merely shifts the costs of its use elsewhere. The wheel allowed crops to be transported and sold a distance from where they were harvested; in doing so, it transformed communities, spreading them out, while also making workers travel farther than ever before in order to make a day's wage. The automobile enabled people to live farther than ever from their loved ones and offices, and our overuse of it has polluted the skies and tarred our lungs. We may "obliterate distance," but another part of nature always pays the price.

We are living in an age of communication revolution—using machines that far outpace human capacity, talking, writing, and typing to one another at greater speed than ever before. We can now work through the night to keep up with colleagues around the globe; we can send an e-mail from New York to Marseilles and have it arrive instantly. We enjoy relatively cheap remote, instantaneous access to our work areas, to newspapers, to stock prices from around the world. "The markets never rest," writes the poet August Kleinzahler in "The Strange Hours That Travelers Keep." "Always they are somewhere in agitation / Pork bellies, titanium, winter wheat."

We are now beginning to understand the consequences of this new circumambient digital "reality." The convenience and speed of the Internet have drawn us powerfully into a virtual world in which distance *appears not to matter*. At the end of the day, though, we need to live in the physical world; we live in communities besides the global village; we need to sleep if we are to live healthfully; we can have five hundred virtual friends, but no matter how often we keep up with them, how many will visit our bedside when we are ill?

We pay the price when we begin to apply the rules of this virtual space to our real life, a lesson that is abundantly clear at work. The Internet was supposed to allow us to work from anywhere, and so we do—from our vacations, from our homes, from our places of worship. It was meant to liberate us from ties to a desk and office space; instead it has led us to work all the time. In the past two decades, we have witnessed one of the greatest breakdowns of the barrier between our work and personal lives since the notion of leisure time emerged in Victorian Britain as a result of the Industrial Age. And we are contributing to this breakdown piecemeal: checking our e-mail in the morning or at night, adding ever more gadgets to our tool kit of connections.

Buying into this way of life—because the Internet is now, at heart, about selling and pitching—has transformed more than just the workplace, though. It has eroded our sense of context, allowing governments and power brokers to create and manipulate a constant state of crisis, because a culture with no sense of its past is blind to the errors of history. It has put us under great physical and mental strain, altering our brain chemistry and daily needs. It has isolated us from the people with whom we live, siphoning us away from real-world places where we gather. It has placed false expectations on real-world living, making us ever more dissatisfied in spite of the huge increase in quality of life in the Western world. It has encouraged flotillas of unnecessary jabbering, making it difficult to tell signal from noise. It has made it more difficult to read slowly and enjoy it, hastening the already declining rates of literacy. It has made it harder to listen and mean it, to be idle and not fidget.

This is not a sustainable way to live. This lifestyle of being constantly *on* causes emotional and physical burnout, workplace meltdowns, and unhappiness. How many of our most joyful memories have been created in front of a screen? Yet in 2006, it was discovered that Americans spent more than half of their life connected to various forms of media. This means we spend more time engaged in media than we do sleeping, more hours plugged in than we log at work. We work in order to have time to watch. *We spend more time with our computers than our spouses.* We check our e-mail more often than we drink water. The culture of Western media encourages us to leave a mark on this planet, to transform ourselves and become the most we can be, to project ourselves outward. Yet participating in and keeping up with these media has given birth to the most politically and culturally passive generation the world has ever seen.

If we are to step off this hurtling machine, we must reassert principles that have been lost in the blur. It is time to launch a manifesto for a *slow communication movement,* a push back against the machines and the forces that encourage us to remain connected to them. Many of the values of the Internet are social improvements—it can be a great platform for solidarity, it rewards curiosity, it enables convenience. This is not the manifesto of a Luddite, this is a human manifesto. If the technology is to be used for the betterment of human life, we must reassert that the Internet and its virtual information space is not a world unto itself but a supplement to our existing world, where the following three statements are self-evident.

1. Speed Matters.

We have numerous technologies that can work with extreme rapidity. But we don't use these capabilities because they are either dangerous (even the Autobahn has begun applying speed limits, due to severe accidents) or uncomfortable (imagine turbulence at 1200 miles per hour) or would ruin the point of having the technology at all (played back faster than it was recorded, Led Zeppelin's syrupy metal sound turns to tinsel).

The speed at which we do something—anything—changes our experience of it. Words and communication are not immune to this fundamental truth. The faster we talk and chat and type over tools such as e-mail and text messages, the more our communication will resemble traveling at great speed. Bumped and jostled, queasy from the constant ocular and muscular adjustments our body must make to keep up, we will live in a constant state of digital jet lag.

This is a disastrous development on many levels. Brain science may suggest that some decisions can be made in the blink of an eye, but not all judgments benefit from a short frame of reference. We need to protect the finite well of our attention if we care about our relationships. We need time in order to properly consider the effect of what we say upon others. We need time in order to grasp the political and professional ramifications of our typed correspondence. We need time to shape and design and filter our words so that we say exactly what we mean. Communicating at great haste hones our utterances down to instincts and impulses that until now have been held back or channeled more carefully.

Continuing in this strobe-lit techno-rave communication environment as it stands will be destructive for businesses. Employees communicating at breakneck speed make mistakes. They forget, cross boundaries that exist for a reason, make sloppy errors, offend clients, spread rumors and gossip that would never travel through offline channels, work well past the point where their contributions are helpful, burn out and break down and then have trouble shutting down and recuperating. The churn produced by this communication lifestyle cannot be sustained. "To perfect things, speed is a unifying force," the race-car driver Michael Schumacher has said. "To imperfect things, speed is a destructive force." No company is perfect, nor is any individual.

It is hard not to blame us for believing otherwise, because the Internet and the global markets it facilitates have bought into a fundamental warping of the actual meaning of speed. Speed used to convey urgency; now we somehow think it means efficiency. One can even see this in the etymology of the word. The earliest recorded use of it as a verb—"to go fast"—dates back to 1300, when horses were the primary mode of moving in haste. By 1569, as the printing press was beginning to remake society, speed was being used to mean "to send forth with quickness." By 1856, in the thick of the Industrial Revolution, when machines and mechanized production and train travel were remaking society yet again, "speed" took on another meaning. It was being used to "increase the work rate of," as in speed up.

There is a paradox here, though. The Internet has provided us with an almost unlimited amount of information, but the speed at which it works—and we work through it—has deprived us of its benefits. We might work at a higher rate, but *this is not working.* We can store a limited amount of information in our brains and have it at our disposal at any one time. Making decisions in this communication brownout, though without complete information, we go to war hastily, go to meetings unprepared, and build relationships on the scree of false impressions. Attention is one of the most valuable modern resources. If we waste it on frivolous communication, we will have nothing left when we really need it.

Everything we say needn't travel at the fastest rate possible. The difference between typing an e-mail and writing a letter or memo out by hand is akin to walking on concrete versus strolling on grass. You forget how natural it feels until you do it again. Our time on this earth is limited, the world is vast, and the people we care about or need for our business life to operate will not always live and work nearby; we will always have to communicate over distance. We might as well enjoy it and preserve the space and time to do it in a way that matches the rhythms of our bodies. Continuing to work and type and write at speed, however, will make our communication environment resemble our cities. There will be concrete as far as the eye can see.

2. The Physical World Matters.

> *Here in the United States one can see what technology has done to rural areas. The small community of a hundred years ago, while it has not vanished, is becoming more and more rare as young people go to towns and cities, where there are work and action.*
>
> —EDWARD T. HALL, *Beyond Culture*

The longer we work at speed over virtual forms of communication, the harder it will become to maintain real-world meeting places. Some electronic communication—political organizing, party planning, setting up meetings of any kind, browsing Google's book archive for research—leads to interactions in the tangible here and now. All these activities must be transported to physical-world venues. But a large part of electronic communication leads us away from the physical world. Our cafés, post offices, parks, cinemas, town centers, main streets, and community meeting halls have suffered as a result of this development. They are beginning to resemble the tidy and lonely bedroom commuter towns created by the expansion of the American interstate system. Sitting in the modern coffee shop, you don't hear the murmur or rise and fall of conversation but the continuous, insectlike patter of typing. The disuse of real-world commons drives people back into the virtual world, causing a feedback cycle that leads to an ever-deepening isolation and neglect of the tangible commons.

This is a terrible loss. We may rely heavily on the Internet, but we cannot touch it, taste it, or experience the indescribable feeling of togetherness that one gleans from face-to-face interaction, from the reassuring sensation of being among a crowd of one's neighbors. Seeing one another in these situations reinforces the importance of sharing resources, of working together, of balancing our own needs with those of others. Online, these values become notions that are much more easily suspended to further our own self-interest. Not surprisingly, political movements that begin online must have a real-world component; otherwise they evaporate and dissolve into the blur of other activities.

It is in the interest of large consumer businesses to continue this erosion of the physical commons, however. Storefronts are expensive; retail space can sit empty. Warehoused stock loses value and costs money. Consumers who can be redirected online are effectively helping companies do business in the cheapest way possible, in an environment where they can be pitched to and advertised to more aggressively than ever before. It is almost impossible to navigate the Web without having to stutter-step around ads and blinking messages from sponsors.

In using this tool so heavily, consumers aren't just frying

their attention spans, they're forfeiting one of the large sources of information that comes from face-to-face interaction and business. A butcher can tell you which cuts of meat are the freshest; an online grocer may not. That same butcher, if he is good, might not just remember your preferences—which an online retailer can do frighteningly well—but ask you how your mother has been doing, whether you caught the latest football game. These interactions remind us that we are more than consumers; they remind us that we are part of the world in a way no amount of online shopping ever will.

The Internet's so-called global village is a consumerist dream—ultimate choice, the ability to shop for the lowest price—but, except for the rare tool that redirects back to local businesses, it erodes our ability to get the things we want locally and uses up precious resources while doing so. The Slow Food movement recognized this twenty years ago, when delegates from fifteen countries drafted a manifesto. "In the name of productivity, *Fast Life* has changed our way of being and threatens our environment and our landscapes," they wrote. In other words, we may be able to get oranges from Chile and water from Switzerland, but the carbon emissions involved in shipping them to our doorstep so we can enjoy them are destroying our environment and putting local growers and farmers out of business.

Communication works the same way. If we spend our evening online trading short messages over Facebook with friends thousands of miles away rather than going to our local pub or park with a friend, we are effectively withdrawing from the people we could turn to for solace, humor, and friendship, not to mention the places we could go to do this. We trade the complicated reality of friendship for its vacuum-packed idea. We exchange the real sensual pleasure of sharing a meal or going for a walk—activities that sustain the tangible commons—for the disembodied excitement of being talked to or heard online. Sitting at an outdoor café and having a conversation, browsing for books with a friend in a bookstore, we cannot help but confront the physical world and help maintain its upkeep; chatting online, we can go hours without remembering we are looking into a screen.

3. Context Matters.

Relying on screens, on typing at high speed, we have constructed an environment in which it is more difficult than ever to get a sense of context. We may now be able to shop around for news, but in doing so we have made it harder for one news source to bring it to us effectively. We can chat and correspond faster than ever with colleagues down the hall, but these virtual exchanges tell us far less than a phone conversation or in-person debate. We can receive love letters and licentious gossip faster than ever, but the rate at which such missives come in, followed by yet more correspondence, deprives us of the necessary mental space in which to properly frame our response.

Sitting at the center of it all, our inbox filling up by the hour, makes us feel in control, but the way it has shrunk our frame of reference leads to an ever-widening cultural passivity. We comment glibly rather than engage; there just isn't time. We dial out of news cycles, because there will be a new story tomorrow. We check in with friends in short text messages about inane topics rather than sit down for a proper chat or withdraw to write a letter that can impart thoughts and emotions and give us a sense of our tangible selves in our handwriting, in our choice of stamp, that even the most elegantly composed e-mail will lack.

We need context in order to live, and if the environment

of electronic communication has stopped providing it, we shouldn't search online for a solution but turn back to the real world and slow down. To do this, we need to uncouple our idea of progress from speed, separate the idea of speed from efficiency, pause and step back enough to realize that efficiency may be good for business and governments but does not always lead to mindfulness and sustainable, rewarding relationships. We are here for a short time on this planet, and reacting to demands on our time by simply speeding up has canceled out many of the benefits of the Internet, which is one of the most fabulous technological inventions ever conceived. We are connected, yes, but we were before, only by gossamer threads that worked more slowly. Slow communication will preserve these threads and our ability to sensibly choose to use faster modes when necessary. It will also preserve our sanity, our families, our relationships, and our ability to find happiness in a world where, in spite of the Internet, saying what we mean is as hard as it ever was. It starts with a simple instruction: Don't send.

7

DON'T SEND

I have been a happy man ever since January 1, 1990, when I no longer had an email address. I'd used email since about 1975, and it seems to me that 15 years of email is plenty for one lifetime.

—Don Knuth, Stanford University

In the next decade, the e-mail onslaught will continue to build and build and then build some more. As more people go online, as companies grow and expand internationally, as the pace of globalization increases, our inboxes will become as snarled as Los Angeles freeways during the morning commute. Fewer and fewer parts of the world will be untouched by wireless Internet access; in the summer of 2008, Delta Air Lines became the latest U.S. carrier to announce it would begin offering Internet access on flights. When you can get wireless access five miles above the earth while traveling at 600 miles per hour, you'll soon be able to get it just about anywhere.

If we don't pause to think about whether we need this tool available to us all the time, it will strangle our workdays like a creeper vine on steroids and keep us tied to our machines and our inboxes right up until we crawl into bed. If we all keep agreeing to be continuously available long beyond the working days, deep into our vacations, and everywhere on the street, it will become harder and harder to step away from the computer or the screen. Those who do will be viewed as eccentrics or, worse, cranky Luddites who simply can't handle the modern world.

There are several things you can do to take back control of your life and your workdays and the mental space that is necessary to mindfulness and happy living. I know, since I have watched it happen. Six months before beginning this book, I was receiving two to three hundred messages a day. I would log on in the morning and watch new e-mail march down my Outlook screen with a small bubble of joy—*I was needed!*—and a mountain of dread: if I didn't respond to these messages, I would offend people, miss out on some key piece of business, add to the ever-increasing backlog of messages that was growing like a mulch pile in leaf season.

Trying to keep up with my e-mail, trying to get out in front of it, though, I made every e-mail mistake there was and invented some new ones. I sent "thanks" messages and cluttered inboxes with forwards; I sent messages without properly reading what I was replying to, creating more e-mail; I sent messages in a charged emotional state and definitely offended recipients; I read messages too quickly and was offended myself; I checked my e-mail late at night and first thing in the morning, shrinking my frame of reference to what came in over the e-mail transom, ruining my sleep, and driving my partner crazy through my constant zombielike attachment to my glowing machine; I mistyped and sent messages to the wrong people, forwarded messages without looking at what was in the thread; I subscribed to far too many Listservs and news feeds, making my inbox a combination of to-do list, mailbox, and newspaper; I tried to coordinate complicated conversations and watched them dissolve into name-calling through lack of face-to-face interaction; I got so much e-mail that messages from the people I love got buried and I never answered them.

I began to trawl the Web and the bookstore, looking for solutions. There are dozens of books on e-mail overload, and many of them have helpful solutions. But none of them expands the focus beyond the past fifteen years, which seems like a massive oversight, given that we have leapfrogged far beyond previous generations' rate of correspondence into a stratosphere that feels beyond human capacity. Furthermore, many of the books I looked at recommended yet more technology to solve the problem of e-mail overload. What I am going to recommend here, then, can be accomplished with very few add-ons and further buy-ins to the current system. It begins, too, with one very simple recommendation.

1. Don't Send

The most important thing you can do to improve the state of your inbox, free up your attention span, and break free of the tyranny of e-mail is not to send an e-mail. As most people now know, e-mail only creates more e-mail, so by stepping away from the messaging treadmill, even if for a moment every day, you instantly dial down the speed of the e-mail messagopolis. Your silence doesn't just affect your inbox, it has a compounded effect on the people to whom you didn't send messages: not getting your message will release some time for them to deal with something besides e-mail; and the less time they spend on e-mail, the less e-mail they will send; and so on.

Pausing for even a minute will also give you a chance to ask yourself several key questions that, if they become routine, will vastly improve your ability to use e-mail effectively. Before you send a message, ask yourself: Is this message essential? Does it need to arrive there instantly? Why am I sending it? What expectations or precedents will it set? If you're sending a message to a friend on vacation, why not send a postcard? You might say less, but the physical object will mean more. If it's a message to let someone know you're just thinking about them, why not pick up the phone and leave your friend a voice mail? There are tones and textures to your voice that words cannot convey.

In *Conversation: The History of a Declining Art,* Stephen Miller describes how, as the habit of visiting people in their homes for meals and get-togethers has declined, our ability to hold interesting and rewarding conversations has deteriorated. Relying on e-mail to maintain friendships will only further this downward spiral, because the medium atrophies our ability to listen in real time, where it's considered inappropriate to simply talk. Instead of sending your pal an e-mail or forwarding him a joke, make a date to meet him for coffee. Invite him over for dinner. It might feel uncomfortable at first, but the conversation you have in person will go a lot farther toward keeping your friendship going than a dozen e-mails and text messages scattered over the course of three months.

Not sending e-mail—by which I mean sending a lot less of it—might be difficult at work, but your coworkers will thank you. Eighty percent of corporate e-mail problems are caused by 1 percent of workers who use it inefficiently; are you in that 1 percent? The survey that turned up this statistic showed that one of the biggest generators of excess mail is a medium-sized message sent to a group of people, which then causes a pinball effect as people chime in and comment, having a virtual discussion. If you have a question or a piece of information that is important enough to warrant such a discussion, pick up the phone: it will save everyone involved a series of short interruptions and mitigate the risk of tonal misunderstandings, which, as we have seen, can impede your ability to collaborate in the future.

Not sending e-mail will also help to break people out of the culture of the endless paper(less) trail. If you're working with several people on a project, do all of them need to know you have viewed a piece of information or completed a small task? Giving in to this impulse to cc and bcc sets an expectation of constant contact that actually prevents workers from getting their individual tasks done. This lesson applies to people who work with clients, too; if you e-mail them about every small detail, they will begin to expect that level of hand-holding all the time, even if it multiplies opportunities for miscommunication and eats up time you could be spending on doing your job. If a client is particularly demanding, batch your feedback so you can send an e-mail twice a week that summarizes a number of developments. If you absolutely must create a paper(less) trail, discuss what you are agreeing to do by phone and then send one short follow-up e-mail summarizing what you agreed upon.

2. Don't Check It First Thing in the Morning or Late at Night

Studies have shown that skipping breakfast deprives us of energy for the rest of the day and alters our metabolism. Eating late at night has a reverse effect and can lead to irregular sleep patterns. What we feed our brain at these times has a similar effect on our minds. In Nicholson Baker's novel *A Box of Matches,* a man starts his day by lighting a match and meditating for the time it takes the flame to burn out, which puts him in a calmer frame of mind. "What you do first thing can influence your whole day," he discovers. "If the first thing you do is stump to the computer in your pajamas to check your e-mail, blinking and plucking your proverbs, you're going to be in a hungry electronic funk all morning," he says. "So don't do it."

Not checking your e-mail first thing will also reinforce a boundary between your work and your private life, which is essential if you want to be fully present in either place. If you check your e-mail before getting to work, you will probably begin to worry about work matters before you actually get there. Checking your e-mail first thing at home doesn't give you a jump on the workday; it just extends it. Sending e-mail before and after office hours has a compounded effect, since it creates an environment in which workers are tacitly expected to check their e-mail at the same time and squeeze more work out of their tired bodies.

E-mailing at night creates the same workaholic cycle, especially if it is an executive or higher-up doing it. Every company has a boss who is infamous for logging on and sending messages at three in the morning. This is an incredibly destructive habit, since it encourages other employees to prove they are working just as hard, by responding when they should be entering their third REM cycle or sending messages on the road, on vacation, over weekends. If you are one of these early-morning e-mailers, simply put your message into a draft and send it during business hours; if the idea or information you wanted to pass along was essential enough to wake you, it will be no less relevant when you wake up from a decent night's sleep.

There are, of course, professions in which this rule will be difficult to follow. Investment bankers who depend on keeping up with foreign markets, reporters who are covering breaking stories, doctors who are operating on a patient in need of a new organ, and political campaign employees who have to keep up with a 24/7 news cycle are just a few examples. They will find it nearly impossible not to check e-mail late at night to coordinate and push projects forward. It's important to note, however, that all of these professions have very high levels of burnout; if you're not in one of these jobs but communicate at their frenzied rates, you are applying a false standard to your work environment.

Just living in the new global workplace isn't a reason to try to keep up the current e-mail pace. Luis Suarez, a social computing evangelist for IBM who lives on the Canary Islands and reports to managers in the United States and Holland, recently decided to wean himself off his e-mail habit. Within a week he found that he had cut back the amount of e-mail in his inbox by 80 percent. He accomplished this by using Web-based free tools such as blogs, where he could post information and make it available to a large group of people; social networking sites, which build trust; and instant messaging, which allows him to interact in real time when it is important. He also just stopped sending a lot of e-mail. Suarez found he got much more done and had more time to enjoy, well, being on the Canary Islands.

3. Check It Twice a Day

Although it is impossible for many of us to imagine, it is possible to check your e-mail far less frequently, even just twice a day, and get more accomplished while doing so. H. L. Mencken checked and answered his mail twice daily and was able to respond to eighty letters because he scheduled the time to do so; there was also a lag time between when his missives went out and when they arrived, and again if he received a response. If they had been e-mails he was sending, his replies would probably have generated yet more correspondence the same day, as some of the eighty recipients would most likely have written back right away.

It is unlikely that the time it takes e-mails to arrive will ever slow down. Still, following Mencken's rule will free up large chunks of your day formerly spent fidgeting with e-mail (and creating yet more of it); even so, your messages will arrive exponentially faster than Mencken's since all of them will get there instantly. You will also be fully present when you get around to sending your messages, rather than fighting between bits and blips of your day. As a result, your e-mail will be more lucid and better designed to get a clean, fast response.

Checking your e-mail twice a day (or even just once an hour) will also allow you to set the agenda for your day, which is essential if you want to stay on task and get things done in a climate of constant communication. How many times do we get so distracted by e-mail that we lose focus and forget there was something very important we had intended to complete during the workday? And then you have to stay late at work, when it's quiet and e-mail isn't coming in as fast, to get it done. Checking your e-mail once in the morning and again in the afternoon will allow you to scroll through messages, pull out the urgent ones, add them to your agenda, and respond accordingly. Your to-do list will remain intact and then shrink.

If you work in an environment where many people keep their inboxes open all day, put an automatic return message on your mailbox directing people to contact you through other channels. It can be a simple message and needn't scold people for using e-mail. WebWorkerDaily cited this message as an example:

> Due to a technical issue, there is a possibility I may never see your email. If it is important, please call me at xxx xxx-xxxx.
>
> Sorry for any inconvenience.

For a brief time, this autoreply will create more messages for your correspondents and contacts, but it will quickly train them out of e-mailing you about small or insignificant things, freeing up time for you to do important work and actually become a more useful colleague or contact. If the message is worded correctly, it will also reinforce your availability without cutting off communication. If their needs are important, your friends and colleagues will get in touch with you and your time away from the e-mail Tilt-a-Whirl will allow you to address them more adequately.

Following this rule is especially important if you have built up a giant backlog of e-mail. Some e-mailers recommend that you simply delete them all and start fresh. This always seems extreme to me. A great many of the backlog e-mails can probably just be deleted, but there will be many that you should probably address and some that might be irreplaceable. If you have amassed a large backlog, set them aside in a file folder in your inbox and budget time every day to deal with them. Make a goal—such as cleaning your inbox backlog in thirty days—and track your progress. Be ruthless about what gets a response. Many of these e-mails will be from people you're currently corresponding with, so you can easily batch your responses right in with current correspondence. If you can manage to check your e-mail less—if not twice a day, then once an hour, tops—you will be creating less e-mail, and that will make it less likely you'll have to face a backlog again.

Staying true to the twice-a-day rule is crucial for anyone

who works in a job that requires sustained concentration. For this reason, writers and journalists the world over have begun to ration their Internet time, pulled Wi-Fi out of their homes and computers, and in some cases started keeping two separate computers, one for writing and the other for communicating with the outside world. It's not just writers, though. Donald Knuth, a computer science professor at Stanford University, has actually given up e-mail altogether. "Email is a wonderful thing for people whose role in life is to be on top of things," he writes on his Web site. "But not for me; my role is to be on the bottom of things. What I do takes long hours of studying and uninterruptible concentration."

4. Keep a Written To-do List and Incorporate E-mail into It

Even if you want to stay on top of things in a fast-paced environment and be the person who appears to do it all, it is essential for you to pull yourself out of the e-mail interface and create a to-do list. Many technology consultants will recommend that you add new applications, more sophisticated gadgets, and new for-pay services that coordinate all your activities, rearrange how you work, and become further enmeshed with a computer program. None of this is necessary. Remember the PalmPilot? It did all that and more, yet, in the end, it was basically an expensive address book and message pad.

For a couple of bucks you can buy a legal pad and keep a rolling to-do list on it. If this is the first interface you check in with every day and the last you consult at the end of the day, you will save yourself a lot of hassle when *urgent* e-mail comes in and threatens to bump you off schedule. After you check your e-mail in the morning, add the messages that came in that cannot be responded to immediately to the list and set a time frame for answering them. If it's going to take a few days to get back to someone, and that person seems like the type of correspondent who might expect an immediate answer, e-mail him or her back to set expectations: "Thanks, Don, it's going to take me a day or two to get to this. I'll e-mail you when I've got an answer." Frantic e-mailing often occurs when people feel as if they haven't been heard. A short reply e-mail at the time a request arrives could save you a few exchanges down the road as you set and rearrange time frames.

Keeping a separate, written to-do list will help pull you out of the computer and into a frame of mind where you can think realistically about what is and isn't important and how long it will take you to do a task. Working in e-mail all day, the batting back and forth, the constant typing gives us the impression that everything can be accomplished in a workday, that it must be finished, or else we have *homework*. No matter how much you love a job, holding yourself to this standard will eventually make you resent it; when six or seven o'clock rolls around, there will always be one more message to write, one more ping to clock in. If you have a written to-do list, however, all those inevitable cross-outs and updates will serve as a reminder that all things change, things need to get done, but timetables are constantly being updated.

5. Give Good E-mail

E-mail can be extremely useful when used well because it has many things a letter does not, starting with the subject line. Use it. If you have a very short message—"I can meet you tomorrow at 5"—that can be stated in a direct, friendly way (you don't want to sound as if you are giving orders), simply put it in the subject line. If you are passing on information that someone else can use but isn't urgent, simply type FYI; to really cut down on your responses, type EOM (end of message) in the subject line of such messages, or "no response needed," both of which will make it clear that you don't expect or really want a "thanks" or a comment. These phrases may feel robotic at first, but using them will help your recipient sort your message and it will ensure that it gets answered in a timely fashion. If something is truly urgent, call the other person; those red exclamation points should be used sparingly, if at all.

Messages should almost always be short. Writing short messages will help your recipient figure out what you need and get back to you quickly. If you have questions, separate them out in the text of the message so that it's clear that you're asking for more than one piece of information. If you have more than a couple of questions, call someone on the telephone or—if he or she works better in chat format—send an instant message to schedule a call in order to minimize the back-and-forth or to work efficiently in an environment where back-and-forth is the norm. If you're sending a message to a friend and it grows to more than a few paragraphs, what you're writing is a letter, not an e-mail. Does the message need to get there within a few minutes? If not and you want to send a gesture, print it out or write it out by hand and send it as a letter. It leaves a small carbon footprint, will be a welcome relief from bills and credit card solicitations, and will stand out even more than if you sent it by e-mail, where it has a chance of getting buried and becoming just one more piece of e-mail on the heaping mound of guilt in many people's inboxes.

E-mail may be fast, but opening and dealing with it takes

time. So don't bludgeon people with serial messages. Setting an e-mail schedule—twice a day, again, is best—will force you to stop and think about what you need from people or what you want to tell them. Working with e-mail constantly open gives the impression that all people are constantly available. Nothing looks worse than sending five, six, seven messages to the same person within a short time frame: "Oh, yeah, I forgot something." This type of machine-gunning makes you look disorganized and will decrease the chances that your e-mails are read closely, since your recipients will start to expect an endless series of updates. *Here we go again.* Eventually, your correspondents will start to sit on your messages so they can batch replies—and if this happens, you have forced your respondent to get organized for the both of you.

Finally, if you want people to contact you through other channels, which is essential if you are going to cut down on your e-mail, make sure that you have a signature on your e-mail that provides your mailing address, your telephone numbers, and the address of a Web site or your Skype address. Sending e-mail without this information is almost like sending a letter without a return address. You are basically ensuring that people have only one way to contact you: by e-mail.

6. Read the Entire Incoming E-mail Before Replying

This seems like a pretty basic rule, but a great deal of e-mail is generated by people replying without having properly read initial messages. If you follow the first five rules, following this one will become second nature. You will have less e-mail to read, and you will be in a more concentrated frame of mind when you are in fact checking it. Following this rule will also bring a bit of civility back into e-mail correspondence. There's nothing worse than answering someone's question thoughtfully and then getting a fast reply that shows they completely ignored all the time you put into crafting a message. If someone has asked you for several pieces of information, answer what you can and set a timetable for when you will get back to that person on everything else. Mark it on a to-do list and then put the initial message in a "pending" folder. Slowly, your inbox will clear out, and you won't feel such a mixture of doom and confusion upon opening it.

If you need to keep a paper trail and you want to save paper, file your outgoing message only; that way you have both the request and the reply. E-mail advisers are of mixed minds about file folders; Mark Hurst, a user experience consultant who has developed an online to-do list, is flatly against it. But the truth of the matter is that important information and communication will continue to arrive by e-mail, and creating a small number of files will make your inbox that much more organized and empty. Think hard about how much you want to file, however; the only thing more annoying than being buried in e-mail is spending most of a day sitting before your computer filing the stuff.

7. Do Not Debate Complex or Sensitive Matters by E-mail

One of the biggest causes of flame wars and e-mail hostility is the idea that we can somehow work effectively while every single one of our visual cues is browned out. We can't. The law of averages is against us. If one of every two messages is misunderstood, working with someone exclusively by e-mail will build up a backdraft of irritation and hurt feelings. This can become a serious situation when you decide to discuss job performance or want to offer constructive feedback, register frustration, apologize, or plead for help. In all such cases, if you can do it, a face-to-face meeting is best. If you can't meet face-to-face, talk by telephone or, better yet, download Skype, buy an inexpensive camera—if you don't have one built into your computer—and have a video chat; it will be cheaper than making a long-distance call, and you will be able to see the other person.

Recognizing the importance of steering clear of sensitive topics in e-mail is especially important at work. Complex questions of strategy or planning that involve people's jobs and livelihoods, their sense of intelligence, requires clear thinking and careful parsing, both of which are almost impossible to do in a climate of frenzy. They're even harder to do in an environment in which all your natural communication receptors are blocked from functioning. Sit down face-to-face, and, as difficult as it may be, your awkwardness in delivering a message—your expressions, your hesitations, your self-interruptions—will go a lot further to convince the person you're talking to that you are aware that he or she is on the receiving end of a message that may be hard to spit out but essential nonetheless.

8. If You Have to Work as a Group by E-mail, Meet Your Correspondents Face-to-Face

In *Here Comes Everyone,* Clay Shirky described how when he was a computer programmer he worked on an important project that involved communicating with people in remote offices by e-mail. Eventually, misunderstandings ground the group nearly to a halt. They decided to fly out to a city and meet in person, and all of the hostility vanished when they met one another face-to-face and realized they were not horrible people. Morale improved, and the team met its deadline.

Many more companies in our era are spread out around the globe or throughout geographic regions, making it difficult for people working together to get together in the same room. Companies that fly people to a city to meet and get together will make an invaluable investment in collective goodwill and long-term productivity, but we're at the stage of climate change where environmental factors should be taken into consideration. Companies need to ask themselves: How essential is a meeting? Who should the key participants be? If people do not work in the same building or city, buy them all Skype or another video chat service so that they can see one another and have something at least vaguely resembling face-to-face interaction. All the hours that are rescued by not having to put out fires, mediate disputes, and patch up miscommunications will be opened up to pick up new business, rest and recuperate, or move on to new projects.

9. Set Up Your Desk to Do Something Else Besides E-mail

In the past two hundred years, human beings have undergone a radical shift in our environment. We have always adapted our environment to our needs—some argue that this is what makes us human—but now we have entered a postnatural world. We have extended day into night, we have developed technologies of movement that allow us to travel far faster than a human or horse can. We have developed numerous forms of technology that allow us to communicate without touching anything or anyone at all.

As much as you can, take control over your office space by

setting aside part of your desk for work that isn't done on the computer. Imagine it as your thinking area, where you can read or take notes or doodle as you work out a problem. The computer screen can sometimes feel like an all-day-long electrical interrogative; it will take as much as you can ask of it and still be ready for more. There are many jobs, however, that would benefit from workers' swiveling away from the screen and concentrating fully on a task. Building your work area up in a way that allows you to do that may save your sanity and make you more efficient.

10. Schedule Media-free Time Every Day

At the end of the day, it is important to remember that we were not born to e-mail, download, watch YouTube, and play online games. We had to learn to do all these things, and if the time we spend during work hours using these tools and toys leaves us agitated, we can also decide not to use them after work. If plugging in at night has become a habit, you might need to remind yourself that there's another way to spend your evening hours. For a week try to shut off all media at nine or ten o'clock, and see what happens to your body and mind. Set the parameters as narrow or as broad as you like, but make sure it means at least no e-mail and messaging. If you want to be bold, cut out all screens—TV, video, handheld, computer.

If you have been having trouble sleeping, it's possible you might gradually feel your body slow down at last. The person who has been sitting next to you on the couch might sharpen in focus; stresses at your job could begin to seem less like Armageddon and more like something happening *over there*. That cluster of what-ifs and niggling doubts that is occluded by work and the pace of your schedule might slide back into view. Maybe you will finally go back to French lessons, read the newspaper; maybe you can make it home in time to watch your daughter's soccer game. You might begin to think of the people in your life who aren't always in touch via e-mail, wonder how they're doing.

This awareness—this presence in the moment—is what gives us the power to act and make decisions, to shape our own lives and truly touch other people. Life might change in the blink of an eye, as the cliché goes, but we also take part in that change and steer it by living purposefully. Of all the things e-mail overload robs from us—be it the pleasure of our work, the strength of our eyes, or even the relationships that flame out through its lack of social cues—this is the most severe. Watching the abuse of e-mail start to choke this mindfulness out of friends and coworkers, strangers on the street, I realized that this is not simply an issue for efficiency experts but a serious epistemological concern for our society. It is affecting our capacity to know one another and the world, to listen. It's grave enough that if we allow it to continue unchecked, we will have nothing worth sending—a terrible loss. Worst of all, if these past few years are any gauge, that won't stop e-mail from arriving, either.

SELECTED BIBLIOGRAPHY

Adams, Henry. *The Education of Henry Adams.* Ed. Ernest Samuels. Boston: Houghton Mifflin, 1973.

Alvarez, A. *The Writer's Voice.* New York and London: W. W. Norton, 2005.

Anderson, Benedict. *Imagined Communities: Reflections on the Origin and Spread of Nationalism.* New York: Verso, 2006.

Baker, Nicholson. *Vox: A Novel.* New York: Random House, 1992.

Bamford, James. *A Pretext for War: 9/11, Iraq, and the Abuse of America's Intelligence Agencies.* New York: Doubleday, 2004.

Baron, Naomi S. *Alphabet to E-mail: How Written English Evolved and Where It's Heading.* London: Routledge, 2001.

———. *Always On: Language in an Online and Mobile World.* New York and London: Oxford University Press, 2008.

Beard, George Miller. *American Nervousness: Its Causes and Consequences.* New York: Putnam, 1881.

Berger, John. *And our faces, my heart, brief as photos.* New York: Vintage, 1991.

———. *Ways of Seeing.* London: Penguin Books, 1972.

Blaise, Clark. *Time Lord: Sir Sandford Fleming and the Creation of Standard Time.* New York: Vintage, 2000.

Brock, Gerald W. *The Second Information Revolution.* Cambridge, Mass.: Harvard University Press, 2003.

Burroughs, William S. *Word Virus: The William S. Burroughs Reader.* New York: Grove, 2000.

Cixous, Hélène. *Stigmata: Escaping Texts.* Oxfordshire: Routledge, 2005.

Crystal, David. *Language and the Internet.* Cambridge, England: Cambridge University Press, 2001.

Davis, Mike. *Dead Cities.* New York: New Press, 2002.

Debord, Guy. *The Society of the Spectacle.* Brooklyn: Zone Books, 1995.

DeGrandpre, Richard. *Digitopia: The Look of the New Digital You.* New York: AtRandom.com, 2001.

Dos Passos, John. *1919.* New York and Boston: Mariner, 2000.

Emerson, Ralph Waldo. *Selected Writings of Ralph Waldo Emerson*. New York: New American Library, 1965.

Faust, Drew Gilpin. *This Republic of Suffering*. New York: Knopf, 2008.

Franzen, Jonathan. *The Corrections*. New York: Picador, 2002.

Goffman, Erving. *The Presentation of the Self in Everyday Life*. New York: Anchor, 1959.

Hafner, Katie, and Matthew Lyon. *Where Wizards Stay Up Late: The Origins of the Internet*. New York: Simon & Schuster, 2006.

Herodotus. *The Histories*. Trans. Aubrey de Selincourt. New York: Penguin, 2003.

Holub, Miroslav. *The Dimension of the Present Moment and Other Essays*. London: Faber and Faber, 1990.

Homer. *The Odyssey*. Trans. Robert Fagles. New York: Penguin, 2006.

Howe, Daniel Walker. *What Hath God Wrought: The Transformation of America, 1815–1848*. New York and London: Oxford University Press, 2007.

Iacoboni, Marco. *Mirroring People: The New Science of How We Connect with Others*. New York: Farrar, Straus and Giroux, 2008.

Jameson, Fredric. *Postmodernism, or the Logic of Late Capitalism*. Durham, N.C.: Duke University Press, 1991.

Johnson, Steven. *Interface Culture: How New Technology Transforms the Way We Create and Communicate*. New York: Basic Books, 1997.

Kern, Stephen. *The Culture of Time and Space 1880–1918.* Cambridge, Mass.: Harvard University Press, 1983.

Lasch, Christopher. *The Culture of Narcissism: American Life in an Age of Diminishing Expectations.* New York and London: W. W. Norton, 1991.

Lesser, Bill, and Steve Baldwin. *Net Slaves: True Tales of Working the Web.* New York: McGraw-Hill, 2000.

Lubrano, Annteresa. *The Telegraph: How Technology Innovation Caused Social Change.* New York: Garland Publishing, 1997.

Marcus, Gary. *Kluge: The Haphazard Construction of the Human Mind.* Boston and New York: Houghton Mifflin, 2008.

Markoff, John. *What the Dormouse Said: How the 60s Subculture Shaped the Personal Computer Industry.* New York: Viking, 2005.

Maslach, Christina, and Michael P. Leiter. *The Truth About Burnout: How Organizations Cause Personal Stress and What to Do About It.* New York: Jossey-Bass, 1997.

McKibben, Bill. *The Age of Missing Information.* New York: Plume, 1993.

Miller, Henry. *Tropic of Capricorn.* New York: Grove Press, 1994.

Niedzviecki, Hal. *Hello, I'm Special: How Individuality Became the New Conformity*. San Francisco: City Lights, 2006.

Paterson, Don. *The Blind Eye: The Book of Late Advice*. London: Faber and Faber, 2007.

Revell, Donald. *The Art of Attention: A Poet's Eye*. Saint Paul, Minn.: Graywolf Press, 2007.

Rothko, Mark. *The Artist's Reality: Philosophies of Art*. New Haven, Conn., and London: Yale University Press, 2004.

Rykwert, Joseph. *The Seduction of Place: The History and Future of the City*. New York: Vintage, 2002.

Scarry, Elaine. *The Body in Pain: The Making and Unmaking of the World*. New York and Oxford: Oxford University Press, 1985.

Schama, Simon. *Landscape and Memory*. New York: Vintage Books, 1996.

Sheeler, James. *Finale Salute: A Story of Unfinished Lives*. New York: Penguin Press, 2008.

Shirky, Clay. *Here Comes Everybody: The Power of Organizing Without Organizations*. New York: Penguin Press, 2008.

Sontag, Susan. *Illness as Metaphor and AIDS and Its Metaphors*. New York: Picador, 2001.

———. *On Photography*. New York: Picador, 2001.

Stille, Alexander. *The Future of the Past.* New York: Picador, 2003.

Trow, George W. S. *Within the Context of No Context.* New York: Atlantic Monthly Press, 1997.

Twenge, Jean M. *Generation Me: Why Today's Young Americans Are More Confident, Assertive, Entitled—and More Miserable Than Ever Before.* New York: Free Press, 2006.

Zilliacus, Laurin. *From Pillar to Post: The Troubled History of the Mail.* London: Heinemann, 1956.

ACKNOWLEDGMENTS

This book would not exist were it not for the gentle, congenial presence of Colin Robinson, who drew the topic out of me over lunch, thereby proving that face-to-face meetings can be vastly more productive than e-mail ping-pong. His wit and encouragement and sharp edits greatly improved the writing, and I will be forever grateful to him for leading me to believe I could even write a book in the first place.

A work of nonfiction always rests upon the work of writers who have come before, and my debts in this case are substantial. I am not a neuroscientist or computer specialist, but the science and technology reporters of *The New York Times,* whose work I quote and refer to throughout the book, dating back to the nineteenth century, and particularly Katie Hafner, were illuminating and pointed me in many fruitful directions.

Early on in the writing of this book, conversations with Lawrence Joseph clarified my thinking, and I am grateful to him for his synthetic intelligence and for pointing me to Stephen Kern's tremendous *The Culture of Time and Space, 1880–1918.* I was also motivated early on by the work of Tom Standage, Clark

Blaise, Naomi S. Baron, and Don Paterson. I was particularly moved by Ayse and Yasar Kemal, who reminded me of the sensuous purpose of creating an artifact.

I am grateful to Scribner for having such a committed team of optimists standing behind a first book. Thank you to Susan Moldow and Nan Graham for their friendship, good company, and determination to get this right. Brant Rumble has been as patient and steady-handed an editor as I could ever hope for, and I am very grateful for his stewardship. Thank you as well to Kate Bittman for humor and belief.

Thank you, Sarah Burnes, for seeing the potential in this book and for encouraging me to take the long view, for being so constantly unflappable, and for chipping in at all the right moments. Thank you, too, Alison Cohen, Stephanie Cabot, and David Gernert. In London I would take to the field with Arabella and Abner Stein any day.

Thank you to Sigrid Rausing and Eric Abraham for generously giving me time out of the office to publish this book, and thank you to the staff of *Granta* magazine, in particular Ellah Allfrey, Liz Jobey, Simon Willis, Roy Robins, Patrick Ryan, and Emily Greenhouse, for putting up with my absence.

Thank you, Richard and Raine Hermsdorf, for storing my library all those years and keeping my office free; Leslie, for getting me to quit.

I perhaps may never have written this book were it not for the love and support of my family; my father, who spent a good part of ten years following me around the streets of Carmichael with a car full of newspapers, showing me how not to quit. Thank you, Andy, for good cheer and your example; to Tim for strength and brilliance; and especially my mother, whom I dearly wish could read this and know it is for her.

Finally, I have to thank the biggest e-mail crank of them all, this book's muse and sternest devotee, who has embraced earnestness in the face of Englishness, and disorder to a point. I am working on the latter. In the meantime, for you I am most grateful of all.

INDEX